T0310123

Human Bond Communication

Human Bond Communication

The Holy Grail of Holistic Communication
and Immersive Experience

Edited by

Sudhir Dixit
Ramjee Prasad

Registered Offices
John Wiley & Sons, Inc., 111 River Street, Hoboken, NJ 07030, USA

Editorial Office
111 River Street, Hoboken, NJ 07030, USA

For details of our global editorial offices, customer services, and more information about Wiley products visit us at www.wiley.com.

The right of Sudhir Dixit and Ramjee Prasad to be identified as the editors of this work has been asserted in accordance with law.

Library of Congress Cataloging-in-Publication Data

Names: Dixit, Sudhir, editor. | Prasad, Ramjee, editor.
Title: Human bond communication : the holy grail of holistic communication and immersive experience / edited by Sudhir Dixit, Ramjee Prasad.
Description: Hoboken, NJ : John Wiley & Sons, 2017. | Includes bibliographical references and index.
Identifiers: LCCN 2016046268 (print) | LCCN 2017002987 (ebook) | ISBN 9781119341338 (cloth) | ISBN 9781119341468 (pdf) | ISBN 9781119341413 (epub)
Subjects: LCSH: Telecommunication systems. | Human-computer interaction. | Digital communications. | Information technology.
Classification: LCC TK5102.5 .H86 2017 (print) | LCC TK5102.5 (ebook) | DDC 621.38201–dc23
LC record available at https://lccn.loc.gov/2016046268

Cover design by Wiley
Cover image: © dem10/gettyimages

Set in 10/12pt Warnock by SPi Global, Pondicherry, India

Printed in United States of America

10   9   8   7   6   5   4   3   2   1

नैव किञ्चित् करमीति युक्तो मन्येत तत्वविद् ।
पश्यन् शृण्वं स्पर्शम् जिघ्रन्नश्नन् गच्छन् स्वपञ् श्वसं ॥

प्रलपं विसृजन् गृहणं उन्मिषण निमिशन् अपि
इंद्रियाणीन्द्रियार्थेषु वर्तन्त इति धारयन् ॥

Naiva kinchit karmiti yukto manyeta tattvit-vit
Pashyan shunvan sparshan jigrhrann asnan gacchan svapan svasan

Pralapan visrijan grihnann unmishan nimishann api
indriyaanindriyaarthesu vartanta iti dhaaryan

One who knows the truth is always certain that it is the senses that are engaged in observations, like seeing, hearing, smelling, touching, and tasting and is the involuntary participant of the actions happening around, just like opening and closing of eyelids. Such observations are not the part of the ultimate knowledge, but, when a seeker looks beyond them, finds the ultimate truth.
                                                        —The Bhagavad Gita (5.8 and 5.9)

Contents

List of Contributors

Ernestina Cianca
Center for Teleinfrastructures
(I-CTIF), University of Rome
"Tor Vergata," Rome, Italy

Maurizia De Bellis
Center for Teleinfrastructures
(I-CTIF), University of Rome
"Tor Vergata," Rome, Italy

Enrico Del Re
Department of Information
Engineering, University of Florence,
Florence, Italy

Mauro De Sanctis
Interdepartmental Center for
Teleinfrastructures (I-CTIF),
University of Rome "Tor Vergata,"
Rome, Italy

Edoardo Di Maggio
I-CTIF Steering Board (LAW-
Intellectual Property), Rome, Italy

Sudhir Dixit
CTIF Global Capsule (CGC),
Rome, Italy
Basic Internet Foundation, Oslo,
Norway

Liljana Gavrilovska
Ss. Cyril and Methodius University in
Skopje, Skopje, Macedonia

Bilal Habib
Wildlife Institute of India, Dehradun,
India

Flemming Hynkemejer
RTX A/S, Wireless Wisdom,
Norresundby, Denmark

Sara Jayousi
Department of Information
Engineering, University of Florence,
Florence, Italy

Geir M. Køien
Faculty of Engineering and Science,
Department of ICT, University of
Agder, Kristiansand, Norway

Pierpaolo Loreti
Interdepartmental Center for
Teleinfrastructures (I-CTIF),
University of Rome "Tor Vergata,"
Rome, Italy

Pradeep K. Mathur
Wildlife Institute of India, Dehradun,
India

Prateek Mathur
CTIF, Aalborg University, Aalborg,
Denmark

Helga E. Melcherts
Varias BVBA, Antwerp, Belgium

Albena Mihovska
Department of Electronic Systems
CTIF, Aalborg University, Aalborg,
Denmark

Seshadri Mohan
Systems Engineering, University of
Arkansas at Little Rock, Little Rock,
AR, USA

Simone Morosi
Department of Information
Engineering, University of Florence,
Florence, Italy

Lorenzo Mucchi
Department of Information
Engineering, University of Florence,
Florence, Italy

Federica Paganelli
CNIT, Research Unit of Florence,
Florence, Italy

Milica Pejanovic
Faculty of Electrical Engineering,
University of Montenegro,
Podgorica, Montenegro

Ramjee Prasad
CTIF Global Capsule (CGC), Rome,
Italy; School of Business and Social
Sciences, Aarhus University, Aarhus,
Denmark

Silvano Pupolin
Department of Information
Engineering, University of Padua,
Padua, Italy

Valentin Rakovic
Ss. Cyril and Methodius University in
Skopje, Skopje, Macedonia

Luca Simone Ronga
CNIT, Research Unit of Florence,
Florence, Italy

Marina Ruggieri
Center for Teleinfrastructures
(I-CTIF), University of Rome
"Tor Vergata," Rome, Italy

Gianpaolo Sannino
Center for Teleinfrastructures
(I-CTIF), University of Rome
"Tor Vergata," Rome, Italy

Sachin Sharma
Systems Engineering, University of
Arkansas at Little Rock, Little Rock,
AR, USA

Domenico Siciliano
Themis Law Firm, Rome, Italy

About the Editors

Dr. Sudhir Dixit recently joined the CTIF Global Capsule (CGC) as the Director of Home for Mind and Body, an international centre for peace, located in Rome, Italy. Additionally, he is a Fellow and Evangelist of basic Internet at the Basic Internet Foundation in Norway. He has also been the CEO and Cofounder of Skydoot, Inc., a start-up at San Francisco Bay area in the content sharing and collaboration space. From December 2013 to April 2015, he was a Distinguished Chief Technologist and CTO of the Communications & Media Services for the Americas region of Hewlett Packard Enterprise Services in Palo Alto, CA, and prior to this he was the Director of Hewlett Packard Labs India from September 2009. From June 2009 to August 2009, he was a Director at HP Labs in Palo Alto. Prior to joining HP Labs Palo Alto, Dixit held a joint appointment with the Centre for Internet Excellence (CIE) and the Centre for Wireless Communications (CWC) at the University of Oulu, Finland. From 1996 to 2008, he held various positions with leading companies, such as with BlackBerry as Senior Director (2008), with Nokia and Nokia Networks in the United States as Senior Research Manager, Nokia Research Fellow, Head of Nokia Research Center (Boston), and Head of Network Technology (USA) (1996–2008). From 1987 to 1996, he was at NYNEX Science and Technology and GTE Laboratories (both now Verizon Communications) as a Staff Director and Principal Research Scientist.

Sudhir Dixit has 21 patents granted by the US PTO and has published over 200 papers and edited, coedited, or authored seven books (Wireless World in 2050 and Beyond: A Window into the Future (2016), Wi-Fi, WiMAX and LTE Multi-hop Mesh Networks by Wiley (2013), Globalization of Mobile and Wireless Communications by Springer (2011), Technologies for Home

Networking by Wiley (2008), Content Networking in the Mobile Internet by Wiley (2004), IP over WDM by Wiley (2003), and Wireless IP and Building the Mobile Internet by Artech House (2002)). He is presently on the editorial boards of *IEEE Spectrum Magazine*, Cambridge University Press Wireless Series, and Springer's *Wireless Personal Communications Journal* and *Central European Journal of Computer Science* (CEJS). He was a Technical Editor of IEEE Communications Magazine (2000–2002 and 2006–2012). He is a two-time winner of the MIT's Technology Review India Grand Challenge Award (2010).

From 2010 to 2012, he was an Adjunct Professor of Computer Science at the University of California, Davis, and, since 2010, he has been a Docent of Broadband Mobile Communications for Emerging Economies at the University of Oulu, Finland. A Life Fellow of the IEEE, and a Fellow of IET and IETE, Dixit received a Ph.D. degree in electronic science and telecommunications from the University of Strathclyde, Glasgow, UK and an M.B.A. from the Florida Institute of Technology, Melbourne, Florida. He received his M.E. degree in Electronics Engineering from Birla Institute of Technology and Science, Pilani, India, and B.E. degree from Maulana Azad National Institute of Technology, Bhopal, India.

Dr. Ramjee Prasad is a Professor in multi-business model and technology innovation in the School of Business and Social Sciences, Aarhus University, Denmark. He is the Founder President of the CTIF Global Capsule (CGC). He has been a Founder Director of Center for TeleInFrastruktur (CTIF) since 2004. He is also the Founder Chairman of the Global ICT Standardisation Forum for India, established in 2009. GISFI has the purpose of increasing the collaboration between European, Indian, Japanese, North-American and other worldwide standardization activities in the area of information and communications technology (ICT) and related application areas.

He was the Founder Chairman of the HERMES Partnership—a network of leading independent European research centers established in 1997, of which he is now the Honorary Chair. He is a Fellow of IEEE (USA), IETE (India), IET (UK), and Wireless World Research Forum (WWRF) and a member of the Netherlands Electronics and Radio Society (NERG) and the Danish Engineering Society (IDA).

He has received Ridderkorset af Dannebrogordenen (Knight of the Dannebrog) in 2010 from the Danish Queen for the internationalization of top-class telecommunication research and education. He has been honored by the University of Rome Tor Vergata, Italy, as a Distinguished Professor of the Department of Clinical Sciences and Translational Medicine on March 15, 2016.

He has received several international awards such as IEEE Communications Society Wireless Communications Technical Committee Recognition Award in 2003 for making contribution in the field of "Personal, Wireless and Mobile Systems and Networks"; Telenor's Research Award in 2005 for impressive merits, both academic and organizational within the field of wireless and personal communication; 2014 IEEE AESS Outstanding Organizational Leadership Award for "Organizational Leadership in developing and globalizing the CTIF (Center for TeleInFrastruktur) Research Network"; and so on.

He is the Founder Editor in Chief of the Springer International Journal on Wireless Personal Communications. He is a member of the editorial board of other renowned international journals including those of River Publishers. Ramjee Prasad is Founder Cochair of the steering committees of many renowned annual international conferences, for example, Wireless Personal Multimedia Communications Symposium (WPMC) and Wireless VITAE and Global Wireless Summit (GWS).

He has published more than 30 books, 1000 plus journal and conference publications, and more than 15 patents and over 100 PhD graduates and a larger number of master's students (over 250). Several of his students are today worldwide telecommunication leaders themselves.

Preface

Applications today have been enriched with multimedia content consisting of audio, video, augmented reality and consistently progressing toward multidimensional rendering, such as stereo, 3D, ultrahigh definition, and fidelity. In parallel, the user interaction with the devices and applications is delivering engaging experience through voice, gestures, gaze, touch, and so on. Wearable devices and body sensors are continually being integrated with applications and user devices, such as a smartphone, remote control, and finding useful applications in healthcare and remote monitoring. Humans interact with applications and consume content through optical and auditory senses. But the understanding is incomplete in the absence of information from and about the other three sensory inputs, namely, olfactory (smell), gustatory (taste), and tactile (touch). This is because all five senses interestingly interact among themselves and the environment, such that being able to sense them, transmit them, and render them at the receiver can potentially deliver powerful experiences. This book on human bond communication (HBC) is about utilizing all five senses to allow more expressive and holistic sensory information exchange through communication techniques for more human sentiment centric communication. The overall outcome is for the human brain to be holistically cognitive of the subject of interest. This complete perceptive information is well exchanged among humans through these senses and, when collectively agreed, becomes knowledge. This is the first book of its kind to motivate research and innovation in holistic communication and to launch a new era of novel products and services to disrupt the status quo of contemporary applications and services that only deal with aural and optical capture, transmission, and rendering of information.

This book focuses on all technologies and issues related to HBC. It also includes the use cases and business opportunities emanating from human-to-machine and machine-to-machine applications, interactions, and communication. The chapters have been authored by the experts in the various fields, which collectively would make HBC possible.

This book is intended for graduate students, academic teachers, scholars, researchers, industry professionals, and software developers interested in the design and development of more engaging and holistic interaction experiences. This book will also be of great interest to casual readers not necessarily familiar with sensor and communication technologies. Therefore, the content is more descriptive and qualitative than theoretical in style of writing.

We thank the contributors of this book for their time and effort to make this book possible in a short period of time. We particularly acknowledge their patience and for always responding promptly to numerous requests for revising their chapters.

Sudhir Dixit
Woodside, California
January 2017

Ramjee Prasad
Aalborg, Denmark

Abbreviations

AAA	Anytime, anywhere, anything
AAL	Ambient Assisted Living
AD	Auxiliary data
AI	Artificial intelligences
ANN	Artificial neural networks
App	Application software to perform tasks for computer/terminal
APT	Advanced persistent threat
APs	Access points
ARPA	Advanced Research Project Agency
ARPANET	Advanced Research Project Agency Network
B2B	Brain to brain
BBI	Brain-to-brain interface
BBU	Broadband unit
BCC	Body channel communication
BCI	Brain-to-computer interface
BIRCH	Balanced iterative reducing and clustering using hierarchies
BMI	Brain–machine interface
ByN	Body as a node
CBD	Convention on Biological Diversity
CBI	Computer-to-brain interface
CCI	Capture, communicate, and instantiate
CDR	Computing device recognition
CERN	Conseil Européen pour la Recherche Nucléaire (European Council for Nuclear Research)
CPS	Calculations per second
CR	Cognitive radio
CRN	Cognitive radio networks
CRNSP	Cognitive radio network service provider
CTIF	Center for TeleInFrastruktur

DARPA	Defense Advanced Research Projects Agency
DBSCAN	Density-based spatial clustering of applications with noise
DNA	Deoxyribonucleic acid
DoD	Department of Defense
DOI	Dolev–Yao intruder
DoS	Denial-of-service
DR	Dead reckoning
DSP	Digital signal processor
DSS	Decision support system
DSLM	Dynamic spectrum leasing methodology/Model
E&Y	Ernst and Young
ECG	Electrocardiogram
EEG	Electroencephalogram/electroencephalography
eHealth	Electronic health
EHR	Electronic health record
EMG	Electromyography
EMR	Electronic medical record
EPC	European Patent Convention
EPO	European Patent Office
EPR	Einstein, Podolsky, and Rosen
ETSI	European Telecommunication Standards Institute
FCN	Fog computing node(s)
FDM	Frequency division multiplexing
fNIRS	Functional near-infrared spectroscopy
FP7	Framework program
FP-growth	Frequent pattern growth
F-RAN	Fog computing-based radio access network
FRAND rates	Friendly, reasonable, and nondiscriminatory rates
F-UE	Fog-capable user equipment
FUS	Focused ultrasound
GDPR	General Data Protection Regulation
GIS	Geographical information system
GPS	Global Positioning System
GSP	Generalized Sequential Pattern
H2H	Human-to-human
H2M	Human-to-machine
HBC	Human bond communication(s)
HBCI	Human bond communication(s) interface
HBP	Human Brain Project
HBS	Human bond sensorium
HCS	Human-centric sensing
HCS-N	Human-centric sensing-network
HCS-NF	Human-centric sensing-network federation

HGP	Human Genome Project
HMI	Human–machine interface
GNSS	Global Navigation Satellite System
HPN	High power node
HPT	Human perceivable transposer
HTML	Hypertext Markup Language
HTTP	Hypertext Transfer Protocol
IAF	Interdisciplinary analysis of functions
IBM	International Business Machines
IC	Integrated circuit
ICN	Information-centric networking
ICT	Information and communication technologies
IdM	Identity management
IEEE	Institute of Electrical and Electronics Engineers
IGW	Internet gateways
IOD	Intraoral device
IoE	Internet of everything
IoH	Internet of humans
IoT	Internet of things
IP	Internet Protocol, intellectual property
IPR	Intellectual property right
IR	Information retrieval
ISDN	Integrated Service Digital Network
KDD	Knowledge Discovery in Databases
k-NN	k-nearest neighbors
LAN	Local area network
LDA	Linear discriminant analysis
LSI	Large-scale integration
METIS	Mobile and wireless communications Enablers for the Twenty-twenty Information Society
ML	Machine learning
M2M	Machine-to-machine
MALDI	Matrix-assisted laser desorption/ionization
MEG	Magnetoencephalography
MEMS	Microelectromechanical systems
mHealth	Mobile health
MMS	Multimedia message(ing) service
MOS	Metal–oxide–semiconductor
MPEG	Moving Picture Experts Group
MRI	Magnetic resonance imaging
mRNA	Messenger RNA
MVNO	Mobile virtual network operator
NFC	Near-field communication

NFV	Network Function Virtualization
NIRS	Near-infrared spectroscopy
NSA	National Security Agency
NLP	Natural language processing
OC	Oral cavity
OCN	Oral cavity as a node
OPTICS	Ordering points to identify the clustering structure
OTO	Old telecom operator
PA	Protected area
PAN	Personal area network
PbD	Privacy by design
PC	Personal computer
PCM	Pulse-code modulation
PH	Partial human
PHR	Personal health record
PI	Pseudoidentifiers
PN	Personal network
PN-F	Personal Network Federation
POTS	Plain old telephone service
POV	Point of view
PrefixSpan	Prefix-projected sequential pattern mining
PU	Primary user
PWA	Physical world augmentation
PWS	Partial wave spectroscopy
QoS	Quality of service
R&D	Research & development
RET	Rare, Endangered, and threatened species
RFID	Radio frequency identification
RNA	Ribonucleic acid
SAR	Structure–activity relationship
S-BAN	Smart body area network
SDN	Software-defined networking
SELDI	Surface-enhanced laser desorption/ionization
SEP	Standard-essential patent
SF	Science fiction
SMS	Short message service
SoC	System on chip
SPADE	Sequential PAttern Discovery using Equivalent Class
STEM	Science, technology, engineering, and mathematics
Stethics	Standardization and ethics
SU	Secondary user
SVM	Support Vector Machine
TCP	Transmission Control Protocol

TFEU	Treaty on the Functioning of the European Union
TMS	Transcranial magnetic stimulation
TPMs	Technological protection measures
TRIPS	(Agreement on) Trade-Related Aspects of Intellectual Property Rights
UHF	Ultrahigh frequency
URI	Uniform Resource Identifier
URL	Uniform Resource Locators
UWB	Ultra-wideband
VANETs	Vehicular Ad-hoc Networks
V2V	Vehicle-to-vehicle
VHF	Very high frequency
VLC	Visible light communications
VM	Virtual machine
VR	Virtual reality
WBAN	Wireless body area network
WSP	Wireless service provider
WWII	World War II

1

Introduction to Human Bond Communication

Sudhir Dixit[1,3] and Ramjee Prasad[1,2]

[1] CTIF Global Capsule (CGC), Rome, Italy
[2] School of Business and Social Sciences, Aarhus University, Aarhus, Denmark
[3] Basic Internet Foundation, Oslo, Norway

1.1 Introduction

Information and communications technologies (ICT) have progressed rapidly in this millennium for people to communicate and exchange information using multimedia (speech, video/image, text), and the same has extended to Internet of things (IoT) and machine-to-machine and machine-to-human communication. This trend is only going to accelerate in the years to come with powerful human–computer interaction technologies to deliver engaging and intuitive experiences. But these developments have remained confined to only the sensing and transmission of aural and optical information in the digital domain through the use of microphone, camera, speaker, and display devices. However, the ability to integrate the other three sensory features, namely, olfactory (smell), gustatory (taste), and tactile (touch) in information transfer and replication to deliver "being there in-person" experience, are still far from reality. Human bond communication (HBC) is a novel concept that incorporates all five sensory information from sensing, to digitization, to transmission and replication at the receiver to allow more expressive, engaging, realistic, and holistic information between humans [1] and in some cases between humans and machines such as in remote sensing and robotic control. Lack of inclusion of the other three senses in the digital world of ICT limits the full exploitation of the cognitive ability of the human mind for a fuller perceptive information experience. The five senses and the environment interact in interesting ways to become complete knowledge for human species as its brain has developed and evolved naturally from the time it came into existence on this planet. The profoundness of perceiving an object depends on the incisiveness and extensity

Human Bond Communication: The Holy Grail of Holistic Communication and Immersive Experience, First Edition. Edited by Sudhir Dixit and Ramjee Prasad.
© 2017 John Wiley & Sons, Inc. Published 2017 by John Wiley & Sons, Inc.

of the sense organs. Incisiveness refers to the granularity and minute details or variations an organ can detect, and extensity refers to the range of the physical property that it can detect.

In the traditional world of digital information exchange, the subject is described and presented partially via its aural and optical rendering, which gives a sense of incompleteness and dissatisfaction in fully understanding the subject. In the present era of ever increasing competition through innovation, inclusion of all five senses to deliver complete experience is the holy grail of the research community. Products have begun to appear through wearables and other embedded sensors in the body, but sensors exploiting touch, taste, and smell and embedding them into products remains a distant reality and is an area of intense research today as would become evident from the chapters included in this book.

Auditory and optical sensing is wave based. In audio sound travels through waves and can be sensed and digitized. Similarly, light shining on an object is reflected in electromagnetic radiation, and a part of this spectrum (called visible light in the range of wavelength 390–700 nm) is visible to the human eye and when rendered on the retina becomes a visual formulation of the object in the nervous system. The camera does this nicely to capture an object visually and digitize it for transmission. When rendered remotely on a display device in 2-D or 3-D, a person can see the object as though he or she was seeing it by being physically present at a location where the camera was located. Other human senses (tactile, olfactory, gustatory) utilize particle-based sensing and rely on smearing the object with the sensors. Building such sensors remains a technological challenge for the research community because each type of sensor must deal with large range of parameters and their wide spectrum. Digitization of these parameters is also a major challenge, and even if some finite widely prevalent values can be captured and digitized, their replication from the digital domain to the analog domain and their sensing by a person in an unobtrusive manner is a complex human-sensor interface issue. Figure 1.1 illustrates the HBC system and depicts what is possible today and what is not.

HBC is about understanding the human sensory functionality and works similar to human sensory system, which includes providing a perceptually holistic understanding of an object combining all five senses while incorporating the object's environment.

1.2   Human Bond Communication (HBC) Architecture

The HBS architecture extrapolates the contemporary communications architecture to include the missing three senses (or types of sensors): tactile, olfactory, and gustatory, not in use today along with the aural and optic

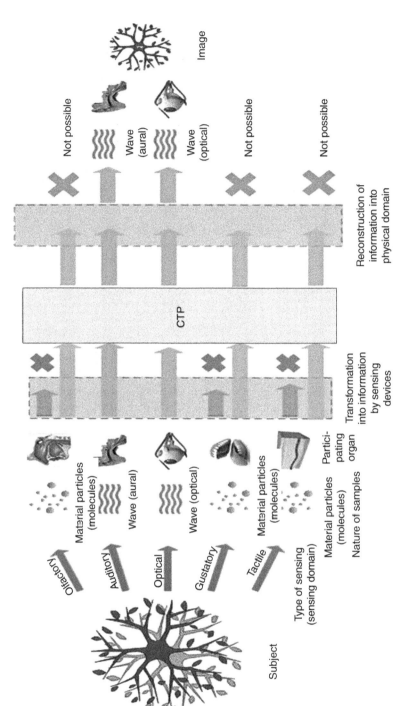

Figure 1.1 An illustration of human bond communication (HBC) concept. CTP, communication technology platform. Prasad [1]. Reproduced with the permission of Springer.

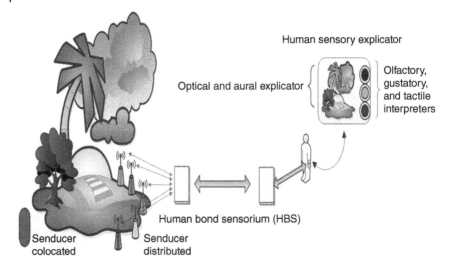

Figure 1.2 A proposed HBC architecture. Prasad [1]. Reproduced with the permission of Springer.

sensors. Nevertheless, some limited deployments are happening in machine-to-machine and machine-to-human communication use cases where robots are being used, such as in industry, law enforcement, hazardous material handling, and surveillance. A proposed architecture is shown in Figure 1.2 [1]. It should be noted that the architecture goes beyond capturing just a person's senses to also deploying all five types of sensors in any environment to capture smell (e.g., types of smoke, air pollutants), tactile information (e.g., surface roughness, temperature, wind speed), and taste (e.g., liquids, dirt, waste) and learning about an object or its surroundings.

The system consists of the three key building blocks: (i) senducers that sense the characteristic parameters through stimuli and transform those analog values to electrical and digital domain for further processing and transmission, (ii) human bond sensorium (HBS) that collects the data from the senducers, processes them to make them consumable for the human perceptive system (i.e., human consumption) by removing a large amount of nonusable and redundant data and information, transmits it to the far end to the receiver gateway, and (iii) human perceivable transposer (HPT) that transforms the received digital data to human consumable format, which includes replication of the senses to a form that one would expect if the person was physically present at the site where the sensory data were collected through senducers. Until such time the replication solutions are not available, the HPT may prefer to render the non-audio–visual sense data through digital means (such as colors, emoticons, text, other gestures like vibration, pressure, temperature, etc.).

1.3 About the Book

Our journey into the world of intuitive and rich communication begins with the vision of extending the contemporary form of digital communication to more natural human-to-human communication through the novel concept of HBC. This chapter has introduced that grand vision. HBC closely embraces the advances in the fields of sensors and wireless distributed computing, physiology, biology, wearables, chemistry, medicine, analytics, Internet, and so on that will be required to bring that vision closer to reality. Therefore, this book has included invited chapters from the experts in the various fields who look at the HBC through their perspectives and delve into the technical challenges that are before the research community. They also discuss the numerous business opportunities that are unlocked due to the intersection of the innovations emanating from interdisciplinary research and entrepreneurship. Whenever appropriate the authors have looked at the historical trends to present their ideas and invoke discourse. Figure 1.3 illustrates some of the key concepts and technologies that will have a profound impact on HBC. These are discussed in the various chapters of the book.

Chapter 1 is an introduction of the book and lays the foundation of the grand vision for the HBC concept.

Chapter 2 presents the basic concepts behind HBC and provides an insight in the ongoing research related to the concepts of human sensory and emotional replication, physical world augmentation, and human umwelt expansion. This chapter then describes an HBC architecture and discusses its convergence

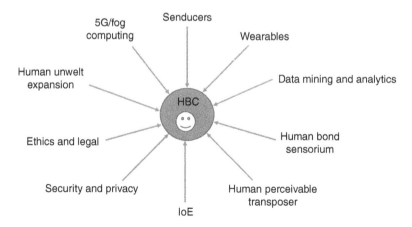

Figure 1.3 Key concepts and technology enablers for HBC.

with ICT. Additionally, the chapter discusses the potentials of HBC and gives a vision of possible future applications and services.

In Chapter 3, the authors postulate that the provision of enhanced augmented reality services to mobile users based on the HBC paradigm will rely on the definition of a high performance, high efficiency, and highly reconfigurable network architecture for the exchange of all the five sensory features. The objective of this chapter is to propose a novel HBC communication network architecture that is able to support the provision of such novel services incorporating all five senses. Starting from the definition of the main network, security, and quality of service requirements for HBC, a 5G network architecture based on software-defined networking, network function virtualization, and Fog–Edge computing paradigms is presented. The main enabling technologies, including WBAN, localization techniques, and content-oriented networking, are described together with some possible solutions to be adopted to cope with the security threats that may affect the success of HBC services.

Chapter 4 is about data mining of the human being. After describing the definition of data mining (also known as knowledge discovery in databases (KDD)) as the process of analyzing data from different perspectives and extracting hidden information and identifying patterns or relationships among the data, the author describes the various models and thereafter focuses on data mining of the human being, where the data is any fact, number, or text regarding a human being. The data can describe the human being at any level, from atoms to cells, to organs, to social level.

Chapter 5 provides an overview of ongoing research on the proposed models for IoT and summarizes their advantages and disadvantages in the context of human centric IoT. After describing potential human centric sensing (HCS) scenarios that require changes in how HCS-based IoT should be modeled, the chapter proposes a macro-level model and describes how it can help to achieve simplicity in the complex IoT world by understanding how to get from micro-complexity to macro-simplicity. It also describes HCS networks and federations and their modeling and later goes into end-to-end security and privacy issues. This chapter also touches upon the concept of tactile Internet as the enabler for HCS IoT.

Chapter 6 describes human body (i.e., body as a node (ByN)) as the main actor in the ICT systems, which plays an active role as a node of the ICT network, as well as part of the ICT user terminal. In addition, "intrusion" with technological ICT devices in the body provides to the body itself a great opportunity for the early monitoring and the daily cure of critical pathologies. After describing the ByN approach, this chapter delves into applying the underlying concept to oral cavity and presents an overview of the research in this field with its implications and perspectives for the future.

Chapter 7 explores the novel machine learning-based approaches to cognitive radio (CR) systems developed that will lead to innovative HBC applications to serve the needs of a community. This chapter formulates novel algorithms to share spectrum through dynamic spectrum leasing methodologies and adaptive policy decision, making processes that seek to maximize the utilization of available scarce spectrum.

Chapter 8 is about the application of ICT for wildlife preservation. It is well known that various governments and nongovernmental organizations have launched diverse technology-driven programs to arrest unprecedented decline and wherever possible successfully restore and rehabilitate wild animal species. While timely integration of technology into wildlife research, monitoring, and conservation in the last couple of decades have definitely yielded positive results, future technology solutions are likely to cater relevant information for decision making and sound management based on application of five human senses instead of just two most common human senses (seeing and hearing). This chapter describes how the sensors for all five senses can be utilized in the solutions for wildlife preservation and concludes that there is an urgent need of sharing mental models between the stakeholders, specifically between the conservationists and technologists.

Chapter 9 investigates the security and privacy issues in HBC. Three different HBC levels are defined and analyzed what these really mean. The approach is to extrapolate and speculate about future progress but to put effort into keeping the extrapolations plausible. Many different fields are involved. Therefore, this chapter serves as a survey about possible future advances in the various fields that will have an impact on HBC. The security and privacy challenges are enormous and they need to be resolved. Thus, this chapter also serves as an urgent call for research in security and privacy issues.

Chapter 10 describes how the Internet of everything (IoE) is the networked connection of people, processes, data, and things. It contains the IoT and the Internet of humans (IoH). The stream of data the IoE will produce can be turned into actionable information and will provide numerous opportunities and will be omnipresent. This chapter attempts to answer the question: Will HBC, the novel concept that incorporates smell, taste, and touch in the exchange of information, be feasible? If the technology to create an HBC ecosystem succeeds, it will bring transformational changes and a paradigm shift. This chapter fast forwards to year 2050 to envision the evolution of the IoE and to predict the anticipated impact and opportunities.

Chapter 11 focuses on the use of HBC for health applications and in particular on the ethical and legal issues that arise. For many years, the use of ICT in medicine was limited to allowing communications between remote patients and doctors (telemedicine). In the recent years, there has been a rapid evolution in the use of ICT in health. The IoT framework allows a pervasive monitoring of anything around and eventually inside us, and this could really open the way to novel diagnostic and therapeutic methods. This rapid evolution has also posed several challenges as many things are not regulated yet.

This chapter attempts to address several key questions: What will happen when HBC will be a reality? Would HBC really enable novel applications in health? And if so, would that require new regulations?

Chapter 12 delves into the challenges in intellectual property (IP) and ICT law that will potentially come with the introduction of HBC. From a legal point of view, HBC means that attorneys and legal professionals should be able to conceive in short time the framework of a smart regulation, in order to provide the principles that will be governing the interaction between human beings, machines, and human umwelt expansion. The opportunities that will be unlocked with HBC will undoubtedly trigger the evolution of IP and ICT regulations in several areas. Because of the need for coherency, a multidisciplinary approach will be the key for reaching consensus among different experts and realize full implementation of the legal and general aspects of HBC.

Chapter 13 presents a historical view of the developments in wireless communication brought about by the changes in paradigm of communications from station to station to person to person and because of technology improvement that made the telephone terminal a multimedia mobile device. This chapter then delves into what is next for wireless communications? While future research could be either on technology or on applications, in reality, the success depends on several other factors such as fashion design, creating user needs, user experience, business models, and so on. These other factors require collaboration among teams in quite different areas that we call for interdisciplinary research and development. This chapter, therefore, focuses on the need for this collaborative approach for innovation and commercial success.

Chapter 14 is a broad overview of how communication among humans originated over the history of mankind and how it has evolved over time with the advances in technology. It discusses the paradox of users that while on the one side they have had choice of platforms and applications to provide enormous opportunities to exchange information in increasingly efficient ways, on the other side they chose the platforms that use only the least significant parts of the messages (i.e., text). This chapter quantifies how much information is included in text, speech, and video/image. Then it discusses technology as an enabler for improving communication over distances and differences between the various platforms, why customers seem not to choose the channel that offers optimum communication, and what are the technical characteristics of the various channels (face to face, letter, telegraph, voice, video, television, SMS/MMS, email, etc.). After presenting the data on how much data the users consume through different channels, this chapter goes into the psychological impact of the various communication channels and finally how the inclusion of the remaining three senses (touch, taste, smell) would further augment the quality of communication.

In summary, the book defines the concept of HBC, sets out its vision, and provides details on the technologies that are driving the realization of the vision and how it would transform the communication experience between humans while also significantly unlock the business opportunities between humans, machines, and their environment. This book also goes into the details of the security, privacy, IP, and regulatory challenges that must be addressed for HBC to be commercially realized.

Reference

1 Prasad, R. (2016). Human bond communication. *Wireless Personal Communications*, **87**(3), 619–627, Springer, New York.

2

General Concepts Behind Human Bond Communications

Liljana Gavrilovska[1], Valentin Rakovic[1], and Sudhir Dixit[2,3]

[1] Ss. Cyril and Methodius University in Skopje, Skopje, Macedonia
[2] CTIF Global Capsule (CGC), Rome, Italy
[3] Basic Internet Foundation, Oslo, Norway

2.1   Introduction

The era of aural and visual communication has been with us for quite a long time, and ICT devices and applications have made tremendous strides in past years by exploiting the exchange of the information associated with these two senses between humans and machines. With respect to aural (or speech and sound) communication, not only have humans exchanged information through vocoders, but also machine-based speech recognition, recording, and synthesis (including through IVRs) are dominating in the present era of digital society. Similarly, visual communication has been growing by leaps and bounds between humans and machines in all forms of combinations. The machines (i.e., robots) are being equipped by voice and video/imaging sensors for machine learning and real-time actuarial functions or for data gathering and forwarding to the destination nodes. Despite all the advances in aural and visual communication, a complete holistic presence, which mimics the face-to-face interaction through ICT, has evaded the technologists so far and constitutes the holy grail of major advances in communication involving human-to-human (H2H), human-to-machine (H2M), and machine-to-machine (M2M) communication and information exchange. Seamless integration of the remaining three senses (touch sensing and haptic feedback commonly referred to tactile system, smell sensing commonly referred to olfactory system, and taste sensing referred to gustatory system) will improve the understanding of the physical object (person, environment, anything else) in a way that we can barely fathom to imagine their impact. The interaction between the five senses

Human Bond Communication: The Holy Grail of Holistic Communication and Immersive Experience, First Edition. Edited by Sudhir Dixit and Ramjee Prasad.

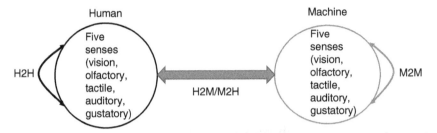

Figure 2.1 Communication of sensory information between humans and machines.

enriches the information content tremendously, which is equivalent to being there. Just imagine how immersive the experience would be from looking at something while touching, hearing, smelling, and sensing its environment remotely! Gustatory experience, while requiring somewhat more active participation, can add to complete experience. A very good example when all the five senses are involved is when we go to a restaurant, to a bar, or on a picnic.

Incorporating the ability to sense all the five senses (or a subset of them) is playing a major role in robotics. Analysis of this five-dimensional data set can provide invaluable information to a robot to learn its environment and how to react, especially in hazardous situations. Alternatively, it can simply pass that information on to its handlers who can then decide on what to do next through that robot remotely or utilize other means to react. Thus, incorporating more than the aural and optical sensory information in the messages is not only limited to H2H communication but also extends to machines and Internet of things (IoT). So far, tactile sensing and feedback (collectively known as haptic or kinesthetic communication) has made significant advances to be implemented in tactile Internet and applications that are remote and based on feel (i.e., cause) and react (control to action) principle [1–3]. Touch conveys such important information as smoothness, texture, shape, pressure, and so on. Examples of such devices are joysticks, data gloves with felt sensors, or other tactile sensors. These are increasingly being used in computer games and remote computer applications, such as remote games and telerobotics. Figure 2.1 illustrates the scope of interactions among five sensory features in H2H, H2M, and M2M communications.

2.2 Definition of Human Bond Communication

The concept of communication of human senses between humans or machines is not necessarily a new concept, but its development and implementation into systems through sensors, how these are packaged for transfer across distances, and how these are rendered are in rather the early stage of implementation and commercialization. It involves interdisciplinary aspects both in theory and implementation in the fields of physics, chemistry, biology, medicine, neuroscience, and engineering. The work mainly entails (i) human sensory and emotional sensing and replication, (ii) physical world augmentation (PWA), and (iii) sensory substitution and augmentation through umwelt expansion. These are explained in the following text [5].

- **Human senses and emotional sensing and replication.** Human sensing and replication typically refers to how humans perceive the world around them through their five senses (including deriving additional intelligence from how these senses interact with each other thereby building a much richer knowledge space in the five-dimensional space). Replication means mimicking and mirroring human senses at the receiving end through artificial means after they have been transmitted digitally. Traditional communication technologies already accomplish this for speech and video involving microphones, cameras, speakers, and display devices for many applications. Human emotional sensing and replication represent a much more advanced concept whereby the brain of a person communicates with another person's brain directly through brain-to-brain (B2B) interface (BBI) comprising brain-to-computer interface (BCI) and computer-to-brain interface (CBI). In recent years, there has been significant progress in research in both the BCI and CBI.
- **PWA.** Interaction with the physical world represented digitally by means of natural human senses represents augmenting the experience that is rich and intuitive. In short, like in the real world, a person utilizes the five natural senses to interact with an object to take appropriate actions and decisions; similarly the same is done in the digital world. The physical world is presented digitally on a device and the user interacts with it through the natural human senses. A good example of this is a wearable device, called "Sixth Sense" device, developed by the researchers at the Massachusetts Institute of Technology in 2009.
- **Sensory substitution and augmentation through umwelt expansion.** The term "umwelt" was first introduced by Jacob von Uexküll [6] and is explained in much more detail in Section 2.7. In brief, they discovered that various organisms in the same ecosystem pick up different signals from the surroundings in the environment they are in. The same applies for the humans as well and thus represents the entire objective reality that goes much beyond the five senses. These environmental signals are fed through unusual sensory channels and together open up interesting interaction possibilities and opportunities between the digital world and the human brain.

Human bond communication (HBC) as a concept and terminology was first introduced by Prasad [7] as a holistic approach to describe and transmit the features of a subject in the way humans perceive it; therefore, all five senses need to be considered and exploited, transmitted, and recreated for complete understanding of the subject to the extent possible by a human and as mutually agreed by the interacting users of HBC. Thus, HBC's scope is end-to-end communication, which includes both humans and machines as end points.

2.3 HBC Architecture and Convergence with ICT

Figure 2.2 illustrates a simplified systems view of an end-to-end HBC system. HBC represents the subsystem that multiplexes or demultiplexes the digital data streams corresponding to the various sensory information, network protocols, and network interface to the external world. In H2M HBC systems the communication of information is usually asymmetric with more sense information flowing from a machine to a user but only control information (such as touch, speech) flowing from user to the machine. To be successful in HBC, the system must perform a number of functions with minimum latency to deliver a real-time immersive experience. These are shown in Figure 2.3.

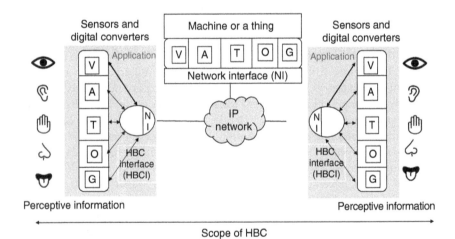

Figure 2.2 A systems level representation of the scope of human bond communication (HBC) system.

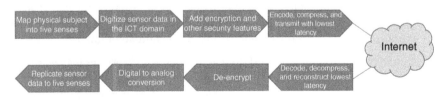

Figure 2.3 Steps to successful implementation of an HBC system.

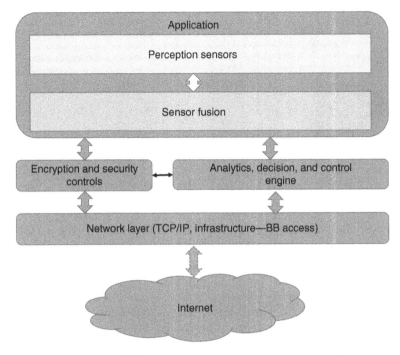

Figure 2.4 A proposed HBC platform with a multisensory user application or a machine.

Schematic representation of an HBC-enabled user and/or a machine (such as a robotic platform) is depicted in Figure 2.4. A complementary stack is implemented in the receiver subsystem of an end-to-end HBC implementation. The communication part of the HBC system remains unchanged as in the conventional multimedia-enabled applications. The recent advances in virtualization and cloud computing can be extended to multisensory applications involving sensing beyond just audio and visual for signal processing, analytics, measurement, and control.

2.4 Human Emotional Messaging

One of the main objectives of HBC is to facilitate the possibility to exchange information on how humans perceive the world. The HBC paradigm exploits the *emotional messaging* concept in order to leverage this process. This section will provide an overview of the key features and technical enablers related to the emotional messaging, highlighting their main research focus and generic issues.

Human emotional messaging represents a technological concept that provides the humans with the possibility to send words, images, and even thoughts

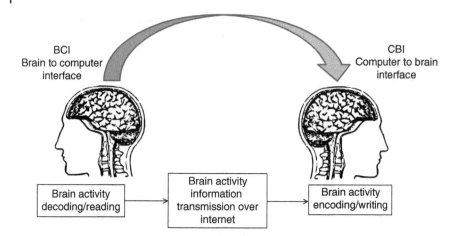

Figure 2.5 Brain-to-brain communication.

directly to the minds of others. The common principle of human information exchange has been supported by the sensory (vision and hear) and motor facets of the human body, which are inept of facilitating the full potential of the emotional messaging. Novel and cutting edge communication concepts, such as the **B2B communication**, facilitate the emotional messaging. As demonstrated in Refs. [8, 9], a direct B2B communication can be achieved by exploiting a *BBI*. The BBI is consisted of two distinct interfaces, as illustrated in Figure 2.5:

1) **BCI.** The BCI is designed to read (or decode) useful information from neural activity.
2) **CBI.** The CBI facilitates the writing (or encoding) of digital information back into neural activity.

Recent research activities have experienced substantial progress in both BCI and CBI communication directions. Specifically several research groups have demonstrated the possibility of decoding motor [9], conceptual [11], and visual [10, 12] information from neural activity via a range of recording techniques such as:

- **Implanted electrodes.** The implanted electrodes are implanted directly into the grey matter of the brain during neurosurgery. Because they lie in the grey matter, invasive devices produce the highest quality of brain activity signals with respect to the BCI procedures. However, they are prone to scar tissue buildup, causing the signal to become weaker, or even nonexistent, as the body reacts to a foreign object in the brain [13].
- **Functional near-infrared spectroscopy (fNIRS).** The fNIRS method exploits the near-infrared spectroscopy (NIRS) for the purpose of functional

neuroimaging. By utilizing fNIRS, the brain activity is measured through hemodynamic responses associated with neuron behavior. With respect to the BCI, the fNIRS method provides the possibility for extracting brain information with high spatial resolution [14]. However, the method lacks good temporal resolution and has a bulky design.

- **Electroencephalography (EEG).** The EEG method is a noninvasive electro-physiological monitoring method to record electrical activity of the brain. It measures voltage fluctuations resulting from ionic current within the neurons of the brain. The EEG method can leverage extraction of brain information with high temporal resolution. The EEG devices can be easily designed to be portable and wearable [15]. However, EEG has low spatial resolution and requires a careful placement of electrodes on head.

- **Functional magnetic resonance imaging (MRI).** The functional MRI represents a functional neuroimaging method using MRI technology that measures brain activity by detecting changes associated with blood flow. This method relies on the fact that cerebral blood flow and neuronal activation are coupled. Similar to the fNIRS method, the functional MRI facilitates extraction of brain activity and information with high spatial resolution and possesses low temporal resolution and complex hardware design [16].

- **Magnetoencephalography (MEG).** MEG represents a functional neuro-imaging technique for mapping brain activity by recording magnetic fields produced by electrical currents occurring naturally in the brain, using very sensitive magnetometers. The MEG provides significantly improved imaging compared to EEG. However, the method possesses a complex hardware design [17, 18].

Varieties of existing CBI techniques permit users to encode digital information into human neural activity by exploiting the facets of implanted electrodes [13], transcranial magnetic stimulation (TMS) [19], and focused ultrasound (FUS) [20]. However, all of these CBI techniques are either invasive or still in an experimental phase. The most recent research advancements in the area of B2B communication provide solutions for noninvasive BBI that can be safely applied to humans. The authors in Refs. [21] and [22] have demonstrated a BBI design, where the BCI exploits the EEG technique to decode motor activities from the *originating* brain and transmit the information via Internet. The information is afterward received and used as input signal to the CBI, based on the TMS technique, and encodes an equivalent activity to the motor cortex of the *receiving* brain. This BBI design facilitates the receiver to interpret the transmitter's command in a remote and "telepathical" fashion.

The B2B communication research activities in the recent years have provided significant results. However, there are still many open issues that have to be addressed in the upcoming years. Specifically, the current B2B communication designs, regardless of the BCI and CBI technology in use, facilitate an information exchange of one bit. This low data transmission rate is a result of the incapability to decode and encode larger volumes of information from and into the human brain. Future research will focus on improving the data rate between the remote communication entities, that is, human brains. Moreover, the B2B communication aspect will open novel possibilities for H2H interrelation, devising many ethical and social implications, thus requiring new legislative actions [23].

2.5 Replication and Translation of Human Senses to Electronic Messages for Communication

Another HBC concept that facilitates the possibility to exchange information on how humans perceive the world is the replication and **translation of human senses into electronic messages**. The HBC paradigm exploits the emotional messaging concept in order to leverage this process. This section will provide an overview of the key features and technical enablers related to the translation of human senses into electronic messages for communication. This section will also highlight the main advantages of the elaborated technology enablers as well as the common issues.

Human sensory replication focuses on digitizing the physical world, based on the input information from the five human senses (i.e., sight, smell, taste, touch, and hearing). It also focuses on replicating the same physical objects based on the existing data, in the digital domain, which portray them. Contemporary communication technologies already incorporate a wide plethora of services and applications that replicate the aspects of sight and hearing. The cutting edge research trends in HBC mainly focus on designing and developing sensor technology that focuses on the other three human senses, that is, touch, smell, and taste. Table 2.1 provides information regarding the digital information rates required to digitize the human senses and include them in the communication loop.

Table 2.1 Human senses and information rates.

Sense	Information rate (bps)
Vision	10^6–10^9
Touch	$<10^7$
Hearing	$<10^7$
Smell	$<10^7$
Taste	$\sim 10^5$

2.5.1 Replication and Translation of Human Touch

The research activities primarily design sensory devices capable of obtaining tactile information through physical contact. The research methodology is also known as *tactile sensing*, where the sensed physical characteristics can be properties such as softness, texture, shape, composition, vibration pattern, temperature, and so on. The research in the area of tactile sensors focuses on several different transduction techniques such as capacitive, piezoresistive, inductive, piezoelectric, strain-based, and optoelectric methods. The intrinsic principles associated with these techniques have their own advantages and disadvantages, which are elaborated in details in Ref. [25]. The capacitive, piezoresistive, piezoelectric, inductive, and optoelectric methods are most commonly utilized in the sensor design, due to their superior performance. Table 2.2 provides an overview of the performance of the different SoA transduction techniques used in tactile sensing.

The tactile sensing research also focuses on restoring sense of replicating both tactile and force information (i.e., haptic feedback). The requirement for tactile replication has increased significantly in recent years, as a result of the development of telerobotics and minimally invasive surgery systems. Similarly, advancements in virtual reality systems have paved the way for the development of tactile gloves that provide force feedback to the fingertips and palms as users touch objects in the virtual environment [1]. The synergy between the tactile sensing, Table 2.2, and the haptic feedback, Refs. [1, 2], can provide the possibility of sending tactile information between remote communication entities. More specifically, it will enable the receiver entity to "feel" a given physical object as perceived by the transmitter entity. With respect to the haptic feedback and tactile replication, most of the presented works utilize the *vibrotactile* actuation. The vibrotactile actuation offers system design that is wearable, lighter, and more energy efficient compared with other methods. However, recent research advancements have shown that novel techniques and equipment that use *air-actuating* system can provide higher degree of realism with respect to the haptic communication [3].

Figure 2.6 depicts a wearable air-actuating system that facilitates a physical interaction in remote communication between parent and child. On the left-hand side of the figure, an input device acts as a tactile interface that allows parents to hug their child and send mood-related expressions. On the right-hand side of the figure, connected through the Internet, an air-actuating module replicates the hug sensation and thus connects the parent and the child via legacy ICT infrastructure.

2.5.2 Replication and Translation of Human Smell

The sense of smell is one of the key human senses. It can provide us with information whether a specific food or beverage is safe for consuming or a

Table 2.2 Comparison of transduction techniques for human tactile sensing.

Transduction technique	Method of operation	Advantages	Disadvantages
Capacitive [26–29]	The technique is based on capacitive coupling that can detect and measure any discrepancies originating from change in conductivity or dielectric properties	High sensitivity Good spatial resolution High dynamic range	Stray capacitance Noise susceptible Design complexity
Piezoresistive [30–32]	The technique utilizes the change in the electrical resistivity of a semiconductor or metal when a mechanical strain is applied	High spatial resolution High scanning rate	Lower repeatability Higher power consumption
Piezoelectric [33–36]	The principle of operation is that a physical dimension is transformed into a force and acts on two opposing faces of the sensing element. Depending on the design of the device, three different detection modes can be used, that is, longitudinal, transversal, or shear	High sensitivity High frequency response High dynamic range	Poor spatial resolution Dynamic sensing only
Inductive linear variable differential transformer [37, 38]	The technique measures any linear displacement position and relies on electromagnetic coupling. It converts a position/linear displacement from a mechanical reference (zero or null position) into a proportional electrical signal-containing phase (for direction) and amplitude (for distance) information	Linear output Unidirectional measurement High dynamic range	Moving parts Low spatial resolution Bulk design
Optoelectric [39–41]	The technique is based on the principle for that the device/sensor of interest produces an electrical signal proportional to the amount of light incident on its active area	Good sensing range Reliability High spatial resolution	Nonconformable Bulk design
Strain gauges [42, 43]	The technique takes advantage of the physical property of electrical conductance and its dependence on the conductor's geometry, when stretched or compressed	Good sensing range Sensitivity Low cost	Calibration requirement Susceptible to temperature changes Design complexity
Multicomponent sensors [44–47]	The technique exploits a combination from the aforementioned principles providing a hybrid solution	Combination of intrinsic parameters	Discrete assembly Higher assembly costs

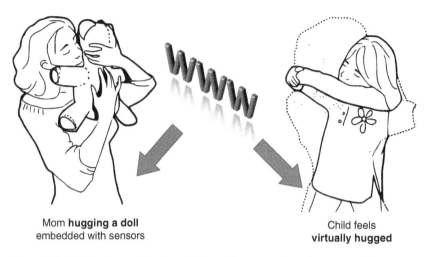

Mom **hugging a doll**
embedded with sensors

Child feels
virtually hugged

Figure 2.6 High-level operation description of the remote hugging system [3].

hazardous situation is occurring (e.g., fire). The olfactory (i.e., smell-based) memory trails an altered time span compared with verbal memory and has a longer lasting effect. Moreover, the olfactory memories have a significantly stronger emotional impact than those that are caused by the rest of the human sensory modalities. Olfactory memory is considered to have reliable qualities, commonly known as "Proustian characteristics," which include resistance to interference, uniqueness, and independence from other modalities [49]. According to the research conducted in Ref. [50], odors can be utilized as means for manipulating mood, decrease stress, improve vigilance, and improve retention and recall of learned subjects. Despite the importance of olfactory information in real-world use cases, until recently it has been almost completely ignored in terms of digitization and replication. The research advancements in this field commonly focus on designing and developing sensory technology that can detect and digitize different aroma types. Usually the olfactory sensors are based on conducting polymer, metal–oxide–semiconductor (MOS), or surface acoustic wave transducers [50–52] and are tailored to digitize specific types of aromas that are commonly found in beverages [53, 54], grains [55], perfume [56], and so on.

In recent years the focus has been steered toward the design and the development of devices that can replicate different types of olfactory aromas. The *oPhone Duo* [57] represents an olfactory device that binds a specific aroma with a predefined image. The device is consisted of an odor-spurting hardware and a mobile application that leverages the users to pair and send photo messages with different scents. In Ref. [4] the authors present a novel mobile

Figure 2.7 Mobile scent actuator. Side view (top left), top view (top right), and separated top view of main unit and scent cartridge (bottom) [4].

scent actuator that can be attached to a mobile phone, for both HBC research and food design. The main focus of the work presented in Ref. [4] is to determine the users' emotional perception of digital images when modulated by scent. This can be an important factor for user experience that could ultimately enhance digital communication and concept of HBC by introducing the olfactory aspect in the communication loop. Figure 2.7 depicts the hardware design of the mobile scent actuator.

Until now there has been no direct relation or demonstration of the combined effect between the smell-related sensors and the aroma replicators. However, proofs of concept solutions such as the ones presented in Refs. [57] and [4] show that in the near future, it can be expected that the synergy of both technologies can introduce the aspect of aroma and smell in the communication loop and provide more immersive communication between remote entities.

2.5.3 Replication and Translation of Human Taste

With similarity to the translation of human smell, the ongoing research in the area mostly focuses on the design and development of sensory technology capable of detecting and digitizing different types of taste. The process of digitization is based on methods that utilize potentiometric–amperometric chemical sensors in combination with the same pattern recognition techniques used for the smell replication technology [59–62] and can be used for digitizing taste such as sweet, bitter, fruity, sour, caramel, artificial, and so on.

The taste replication represents a fairly complex process, where commonly a chain of chemical processes stimulates the sensation of taste. Specifically, the method proposed in Ref. [63] uses chemical and mechanical linkages to simulate food-chewing sensations by providing flavoring chemicals, biting force, chewing sound, and vibration to the user. In Ref. [64] the authors propose a system that sprays chemicals into a wearer's mouth to create different taste sensations. However, the system is considerably large in size and uses different arrays of chemicals to stimulate smell and taste senses. In addition, refilling, cleaning, and durability are several additional weak points of the system. The technology for actuating the human sense of taste with nonchemical methods is still in its early stages. One of the most advanced methods based on nonchemical methods exploits a combination of multiple artificial stimuli [65]. It is based on stimulating the tip of the human tongue through a pair of silver electrodes using electrical and thermal stimulation to simulate primary taste sensations. The system brings these two methods together into a single device and provides a digital control through an attached computer.

The human sensory replication represents a dynamic and ever-growing research area. There exist many research activities that have significantly advanced the process digitizing different types of smell, taste, and touch. However, the main issue with respect to the human sensory replication is the lack of development of sophisticated sensory replicators and communication frameworks that can incorporate the human sensory information in the communication loop.

2.6 Physical World Augmentation

The utilization of communication devices such as mobile phones, tablets, and notebooks provides the end users to reach digital content and information in the anytime, anywhere, and anything (AAA) fashion. However, current research trends fail to address the issues that arise from the gap between the use of ICT devices and the human interactions with the physical world. The PWA concept addresses this issue by providing digital information into the real world. PWA allows humans to seamlessly interact with the required information, conveying an integration of the digital data and the real world.

The most advanced and prevailing model with respect to the PWA represents a wearable device design, developed by researchers at the Massachusetts Institute of Technology, that is commonly used as the basis and standard for the development of the PWA technology. The device is also known as the "Sixth Sense" [66–70] and presents a multisensory platform that facilitates the PWA. The following subsection will elaborate the design, technology, and applicability of the Sixth Sense device in relation to the PWA concept.

2.6.1 The Sixth Sense Device

The Sixth Sense represents a wearable gestural interface device that augments the physical world with digital information. It facilitates the process for people to use natural hand gestures in order to interact with the underlying information. The Sixth Sense design incorporates a mobile communication/computing device, camera, projector, color markers, microphone, and mirror. The camera and projector are connected to the mobile communication device, which utilizes legacy wireless connectivity (e.g., 3G, 4G, Wi-Fi) to access the Internet and the cloud-computing infrastructure that is responsible for the data processing; see Figure 2.8. The whole system is made wearable using a string that holds the projector over the user's chest. The mirror is then attached to the front of a projector. The mirror reflects the projection of visual information to any surface. The core functionalities of the Sixth Sense are based upon gesture recognition and computer vision algorithms. These algorithms facilitate the process for understanding and interpreting the gestures and can identify the physical objects of interest through the device's camera.

2.6.2 Technology Enablers and Use Cases

The PWA concept, as well as the Sixth Sense, relies on the set of technology enablers that facilitate the operation of advanced solutions that are capable of conveying a seamless integration of the digital information and the real world, such as *ubiquitous computing, augmented reality, gesture recognition,* and *computer vision.* Table 2.3 provides a generic overview of the elaborated technology enablers.

The PWA-related technology enablers represent vast standalone research areas. The references in Table 2.3 provide a deeper insight into the most recent surveys and overviews of the related areas.

The PWA concept offers limitless potentials for the development of novel and cutting edge applications. Table 2.4 presents a subset of the existing applications that have been already developed as a part of the Sixth Sense project and the potential ones that may have significant impact in the near future.

The PWA represents a novel and rapidly expanding research area. There are still many open issues related to the concept of PWA that have to be addressed

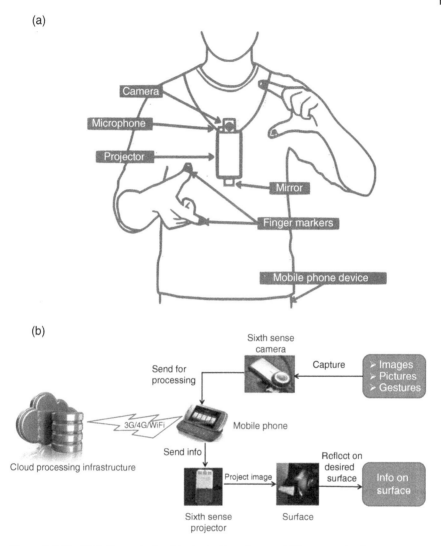

Figure 2.8 The Sixth Sense device: (a) physical topology and (b) communication architecture.

in the upcoming years. For example, the current PWA designs are commonly founded on the Sixth Sense device, thus limiting the scope of applicability of new hardware solutions. However, the open-source nature of the Sixth Sense software framework, as well as the continuous advancements in the research topics elaborated in this section, facilitates the possibility for the extension of the capabilities, applicability, and use cases of the PWA concept.

Table 2.3 Sixth Sense technology enablers.

Technology	Description
Ubiquitous computing	Ubiquitous computing is a concept where computing is made to appear anytime and everywhere. In contrast to desktop computing, ubiquitous computing can occur using any device, in any location, and in any format [71]
Augmented reality	The augmented reality is a visualization technology that allows the user to experience the virtual experience added over real world in real time. Augmented reality adds graphics, sounds, and even haptic feedback to the natural world as it exists [72]
Gesture recognition	Gesture recognition aims at interpreting human gestures with the help of mathematical algorithms. In order to operate, it requires special type of hand gloves, which provide information about hand position orientation and flux of the fingers [73]
Computer vision	Computer vision represents a technology allowing machines to interpret the necessary information from a given image. This technology includes various subfields like image processing, image analysis, and machine vision. It also includes certain aspect of artificial intelligence techniques, such as pattern recognition [74]

Table 2.4 Existing and potential Sixth Sense applications.

PWA application	Description
Capturing photos	By using hand gestures (forming a box), the user can take snapshots from the real world. The box created by the fingers acts as a frame for capturing the photo
Palm as dialer	The technology enables the user to call without using the dialer. The dialer will be projected on the palm, when required by a simple hand gesture and the user can dial the number using the other hand
Interactive reading	By placing a book in front of the user, the device can check for the book's ratings on Internet portals such as Amazon or Google books. Additionally if requested the device can read the book to the user
Dynamic newspaper content	The technology identifies the news headline and then projects relevant textual and video content as additional information to the user
Augmented product information	By placing a product in front of the user, the device can check the content and ratings of the product and thus recommend it or disapprove for purchasing
Access the Internet from anywhere	The users can browse any Internet content, such as www, emails, and so on, on any surface, even on their palm
Real-time language translation	The technology can facilitate users to communicate on their native language and provide real-time translation of the discussion in the preferred language

2.7 Human Umwelt Expansion through Environmental Signals and Context

The term umwelt introduced by Jakob von Uexküll [6] represents a notion that describes a simple and commonly unnoticed observation that different animals in the same ecosystem pick up on different environmental signals. For example, for the tick, the important signals are temperature and the odor of butyric acid; for the black ghost knifefish, the electrical fields; and for the bat, the air-compression waves [75]. Thus the subset of the perceived information from the surrounding ecosystem that any living being is able to detect is denoted as its umwelt. All organisms, including humans, assume that the environmental signals that they exploit represent the entire objective reality. Albeit the human sensory system (i.e., the five human senses) can leverage efficient adaption to the surroundings, it does not approximate all existing environmental data.

The neuroscience research shows that the human brain can interpret and extract the required information from various data streams that are fed through unusual sensory channels. These conclusions have opened the possibilities to explore novel interactions between the available digital data and the human brain and are also known as **human umwelt expansion**.

Specifically, the human umwelt expansion builds on top of two fundamental pillars: *sensory substitution* and *sensory augmentation*. Both of them focus on stimulating the sensory apparatus of an incumbent sense in an attempt to evoke the feel of a novel sense.

For the case of sensory substitution, the novel sense is one that humans with impairments do not possess, for example, hear to deaf personas or sight to blind personas. In the case of sensory augmentation, the new sense is an additional sense, which is not related to the five incumbent human sensory modalities (i.e., sight, smell, taste, touch, and hearing). One example of an application for sensory augmentation is the ability to see in the infrared spectrum. Technology-wise, both the sensory substitution and the sensory augmentation utilize the same sensory stimulation approaches. The sensory stimulation approaches can be divided in two generic classes:

1) Noninvasive approaches
2) Invasive (i.e., bionic) approaches

The rest of the section will elaborate on the most recent research achievements for both approaches.

2.7.1 Noninvasive Sensory Stimulation Approaches

The noninvasive approaches perform the sensory stimulation process (i.e., the transmission of the sensory information) by triggering the incumbent human senses. These approaches focus on either *auditory-* or *tactile*-based transmission of information. In both cases the information of the additional sensory perception is fed to the human brain by noninvasive devices, such as a sound source or vibrating device.

2.7.1.1 Auditory Approaches

The auditory sensory stimulation approaches focus on utilizing the auditory sensory inputs either to compensate for the lack of another sensory input (i.e., sensory substitution) or to augment the existing sensory inputs (i.e., sensory augmentation). The auditory sensory stimulation exploits a range of sensors to detect and store information about the external environment (e.g., normal vision, UV–infrared vision, orientation, balance). This information is then transduced by BCI into auditory signals that are then relayed via the auditory receptors to the brain. Commonly the research activities with respect to the auditory sensory stimulation focus on developing sensory substitution devices that aid the quality of living for humans with impaired sight [76–78]. However, recent research activities have also presented the applicability of auditory sensory stimulation in the case of sensory augmentation for use cases such as targeting, collision awareness, and so on [76, 79, 80].

2.7.1.2 Tactile Approaches

The tactile sensory stimulation approaches exploit the touch-based sensory modality to either compensate or augment the existing sensory inputs. These approaches commonly incorporate vibrating collars or vests for conveying the sensory information to the human brain. The tactile sensory stimulation research stems toward developing sensory substitution devices that aid the quality of living for humans with impaired hearing [58], vision [81], or balance [82, 83]. However, in recent years there has been a significant research performed with respect to the sensory augmentation where the developed tactile devices are used for a variety of applications. The authors in Ref. [84] have proposed a tactile vest for improved performance in precision, whereas the works in Refs. [85] and [24] proposed a tactile feedback design capable of providing improved distance perception. The authors in Refs. [48] and [86] have presented tactile devices that can be exploited for obstacle identification, avoidance, and navigation. In Ref. [87] the authors have proposed a tactile-based system that can be used for improved balance and balance training as well as for improved orientation.

2.7.1.3 Magnetic Approaches

Except for the auditory and tactile approaches, recent research advancements [88, 89] have proposed a novel technological concept for sensory augmentation based on the aspects of electromagnetism. The presented magnetic skin sensory technology is capable of augmenting the human sensory modalities with additional information from the electromagnetic domain. The authors in Ref. [89] also discuss that the proposed technology can be utilized for a variety of use cases with respect to the human umwelt expansion, such as:

- **Magnetoreception**. Navigation and orientation based on the Earth's magnetic poles.
- **Embedded communication**. Decoding electromagnetic "messages" sent by any electronic devices.
- **Functional medical implants**. Monitoring physiological conditions of patients.

Although the magnetic approaches are commonly classified in the noninvasive sensory stimulation group, they can be also included in the invasive group.

2.7.2 Invasive Sensory Stimulation Approaches

The invasive sensory stimulation approaches perform the sensory stimulation by facilitating a direct interface to the human nervous system. Several pioneering works in this area have shown that by using stimulating electrodes implanted into the human nervous system, it is possible to apply current pulses that can be reliably recognized by the user [90]. The applicability of the invasive sensory stimulation approaches is broad and varies from applying tactile feedback to remote robot hands [91] and aiding the auditory process in personas with hearing impairment [92] up to providing a sense of touch to prosthetic hands [93] and partial restoration of sight [94]. Moreover, recent research activities have shown that the invasive sensory stimulation approach can be exploited for expanding the human umwelt with respect to auditory information. Specifically the authors in Ref. [95] have developed a 3-D printed bionic ear that can decode acoustic waves up to a frequency of 100 MHz, as well as decode electromagnetic waves. However, the major obstacle for these approaches is the requirement for invasive procedures that face many legal, ethical, and health and privacy issues [96, 97].

Table 2.5 summarizes the main technological concepts behind the noninvasive and invasive sensory stimulation approaches presented in this section.

Recent research activities related to the human umwelt expansion have significantly advanced the process of providing novel sensory modalities to humans. However, there are still many open issues that have to be addressed. For example, the current noninvasive designs require lengthy training periods in order to leverage the adaptation of the human brain to the new sensory information. Additionally, the invasive designs face many applicability issues due to the requirement of surgical interventions that introduce legal, ethical, and health and privacy issues.

Table 2.5 Comparison of sensory stimulation approaches.

Sensory stimulation approach	Method of operation	Advantages	Disadvantages
Noninvasive approach	Utilize auditory or tactile interface to encode sensory information in the human brain	Low cost Ease of use	Requires training Lower fidelity Bulky design
Invasive approach	Utilize electrode implants connected to the human nervous system to transmit sensory information to the human brain	High fidelity Compact design	Requires surgical intervention Ethical and privacy issues

2.8 Integration of HBC with Wearables and Wireless Body Area Network (WBAN)

Wireless body area network (WBAN) has emerged as an important technology for e-healthcare for health monitoring through wearable sensors and implanted sensors. The sensors can monitor all types of vital parameters, such as blood pressure, heart rate, glucose level, distance walked, analysis of breath, sweat, urine, and sleep. Smartphones have been the key enablers to bring wearable sensors mainstream and affordable (see Figure 2.9). These sensors connect directly either with the smartphone in star configuration or through a gateway function, which may be built in software in the smartphone. Ultra-wideband (UWB) and Bluetooth are the widely used wireless technologies for WBAN and PAN. The various sensors may also be connected in a mesh configuration with the nearest node connecting to the WBAN hub. The sensors are typically tiny and use a very small amount of energy unless they are embedded in wearable device such as a wrist band. Because a person is mobile, the connectivity with the backend system in the cloud is invariably over a cellular data network. Alternatively and in addition, the person himself/herself can view and process the information locally from the smartphone or another similar device. The data collected in the smartphone or in the backend servers at the doctor's office can be analyzed in real time and any abnormalities flagged to the doctor and/or the patient for remedial action. Thus, WBANs are already playing a key role in healthcare monitoring and improving the quality of care.

Adding to these WBAN parameters, the HBC perception and sensation data (such as touch, taste, smell, sound, and sight) and sharing it with someone else remotely would further amplify the experience between humans. As discussed in the previous sections, this would require novel sensors to capture the human sensory information, but once this becomes possible, its communication to a

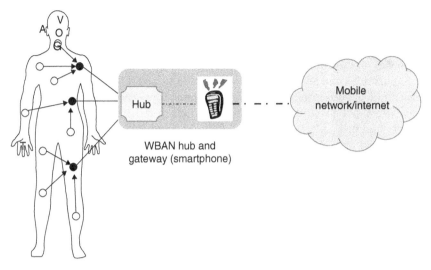

Figure 2.9 Wireless body area network with various medical sensors and human sensory interfaces.

distant location would be no different than transmitting the healthcare data; however, replicating it at the receiving device would require novel synthesizers unless it is initially done through visual means. Capturing of sight and sound is an already well-established process. Another prospective area of research and innovation is how these senses interact with each other and with the environment to convey a total experience of being there. For example, a visually appealing, pleasantly smelling tasty food in a pleasant ambience of sound and scenery can heighten the anticipation and the overall experience of having a meal by an order of magnitude than just sensing one or two sensory stimulants (see Section 2.5.2).

It is important to point out that when considering the e-healthcare applications, the data being generated by the sensors must be accurate all the time and dependable and should reach the receiver in the shortest time possible. The communication from the transmit site (right from the sensors) to the receive site must be highly secured and must render any kind of eavesdropping impossible for the reasons of privacy and data manipulation. Because the body sensors and WBANs must be robust, lightweight, reliable, and energy efficient, the security and privacy solutions can be difficult to develop and implement. Although the HBC data are not health related, they need to be protected similar to healthcare data and should be transmitted in almost real time for an immersive experience, and if stored at the backend servers, they should be accurate for any future analytics and insight to be gained by the service and application provider.

2.9 Conclusions

In this chapter, we discussed the inclusion of three additional human senses to two senses (audio and visual) already being used widely in user and machine/IoT applications and services. When this becomes possible, the world of communication will be enriched to deliver immersive and "being there" experience to the users. We presented the state of the art in detail, innovations on the horizon, and how they can be integrated with the healthcare-related sensors and the environment through WBAN and wearables to unlock significant commercialization opportunities. Exploitation for HBC and other body sensors poses major challenges in the areas of scaling, performance, privacy, and security. How this data is stored, utilized, analyzed, and shared with the interested entities remains an open issue from the regulatory and legal point of view. As with similar challenges faced in the past with the use of the Internet in healthcare, banking, finance, and mission-critical applications, surely these will be addressed and/or sensitivities overcome when traded off with the significant benefits that the users and businesses will experience.

References

1 Hands omni (2015). http://oedk.rice.edu/Sys/PublicProfile/25532450/1063096 (accessed October 10, 2016).
2 Robot shadow hand (2014). http://www.esa.int/spaceinimages/Images/2014/12/Robot_Shadow_Hand (accessed October 10, 2016).
3 Cheok, A., & Pradana, G.A. (2015). Virtual touch. *Scholarpedia*, **10**(4), 32679.
4 Braun, M., Pradana, G., Buchanan, G., Cheok, A., Velasco, C., Spence, C., Aduriz, A., Gross, J., & Lasa, D. (2016). Emotional priming of digital images through mobile telesmell and virtual food. *International Journal of Food Design*, **1**(1), 29–45.
5 Gavrilovska, L., & Rakovic, V. (2016). Human bond communications: generic classification and technology enablers. *Springer Journal on Wireless Personal Communications*, **88**(1), 5–21.
6 Lepschy, G. (1978). Contributions to the doctrine of signs. *Language*, **54**(3), 661–663. http://www.jstor.org/stable/412791.
7 Prasad, R. (2016). Human bond communication. *Springer Journal on Wireless Personal Communications*, **87**(3), 619–627.
8 Pais-Vieira, M., Lebedev, M., Wang, J., Nicolelis, M.A.L., & Kunicki, C. (2013). A brain-to-brain interface for real-time sharing of sensorimotor information. *Scientific Reports*, **3**(1319), 1–10.
9 Li, G., Zhang, D., & Engelmann, J. (2016). Brain-computer interface controlled cyborg: establishing a functional information transfer pathway from human brain to cockroach brain. *PLoS ONE*, **11**(3), e0150667.

10 Nishimoto, S., Vu, A.T., Naselaris, T., Benjamini, Y., Yu, B., & Gallant, J.L. (2011). Reconstructing visual experiences from brain activity evoked by natural movies. *Current Biology*, **21**(19), 1641–1646.

11 Collinger, J.L., Wodlinger, B., Downey, J.E., Wang, W., Tyler-Kabara, E.C., Weber, D.J., McMorland, A.J., Velliste, M., Boninger, M.L., & Schwartz, A.B. (2015). High-performance neuroprosthetic control by an individual with tetraplegia. *The Lancet*, **381**(9866), 557–564.

12 Vondrick, C., Khosla, A., Pirsiavash, H., Malisiewicz, T., Torralba, A. (2016). Visualizing Object Detection Features. *International Journal of Computer Vision*, **119**(2), 145–158.

13 Brennan, J.R., Stabler, E.P., Van Wagenen, S.E., Luh, W.-M., & Hale, J.T. (2016). Abstract linguistic structure correlates with temporal activity during naturalistic comprehension. *Brain and Language*, **157–158**, 81–94.

14 Weyand, S. (2015). Exploring methodological frameworks for a mental task-based near-infrared spectroscopy brain–computer interface. *Journal of Neuroscience Methods*, **254**, 36–45.

15 Fabiani, G., McFarland, D., Wolpaw, J., & Pfurtscheller, G. (2004). Conversion of EEG activity into cursor movement by a brain-computer interface (BCI). *IEEE Transactions on Neural Systems and Rehabilitation Engineering*, **12**(3), 331–338.

16 Yoo, S.S., Fairneny, T., Chen, N.K., Choo, S.E., Panych, L.P., Park, H., Lee, S.Y., & Jolesz, F.A.. (2004). Brain computer interface using fMRI: spatial navigation by thoughts. *NeuroReport*, **15**(10), 1591–1595.

17 Sabra, N., & Wahed, M. (2011). The use of meg-based brain computer interface for classification of wrist movements in four different directions. In: *Radio Science Conference (NRSC), 2011 28th National*, 26–28 April, 2011, pp. 1–7.

18 Perlmutter, J.S., & Mink, J.W. (2006). Deep brain stimulation. *Annual Review of Neuroscience*, **29**(1), 229–257.

19 Hallett, M. (2000). Transcranial magnetic stimulation and the human brain. *Nature*, **406**(6792), 147–150.

20 Yoo, S.S., Bystritsky, A., Lee, J.H., Zhang, Y., Fischer, K., Min, B.K., McDannold, N.J., Pascual-Leone, A., & Jolesz, F.A. (2011). Focused ultrasound modulates region-specific brain activity. *NeuroImage*, **56**(3), 1267–1275.

21 Grau, C., Ginhoux, R., Riera, A., Nguyen, T.L., Chauvat, H., Berg, M., Amengual, J.L., Pascual-Leone, A., & Ruffini, G. (2014). Conscious brain-to-brain communication in humans using non-invasive technologies. *PLoS ONE*, **9**(8), e105225.

22 Rao, R.P.N., Stocco, A., Bryan, M., Sarma, D., Youngquist, T.M., Wu, J., & Prat, C.S. (2014). A direct brain-to-brain interface in humans. *PLoS ONE*, **9**(11), e111332.

23 Klein, E. (2015). Engineering the brain: ethical issues and the introduction of neural devices. *Hastings Center Report*, **45**(6), 26–35.

24 Bertram, C., Evans, M., Javaid, M., Stafford, T., & Prescott, T. (2013). Sensory augmentation with distal touch: the tactile helmet project. In: N. Lepora, A. Mura, H. Krapp, P. Verschure, T. Prescott (eds.) *Biomimetic and Biohybrid Systems*, Lecture Notes in Computer Science, vol. **8064**, pp. 24–35. Springer, Berlin/Heidelberg.

25 Tiwana, M.I., Redmond, S.J., & Lovell, N.H. (2012). A review of tactile sensing technologies with applications in biomedical engineering. *Sensors and Actuators A: Physical*, **179**, 17–31.

26 Wu, X.A., Suresh, S.A., Jiang, H., Ulmen, J.V., Hawkes, E.W., Christensen, D.L., Cutkosky, M.R. (2015). Tactile sensing for gecko-inspired adhesion. Intelligent Robots and Systems (IROS). *IEEE/RSJ International Conference on*, **2015**, 1501–1507.

27 Wei, Y. (2015). An overview of micro-force sensing techniques. *Sensors and Actuators A: Physical*, **234**, 359–374.

28 Peng, J.Y., & Lu, M.C. (2015). A flexible capacitive tactile sensor array with CMOS readout circuits for pulse diagnosis. *Sensors Journal, IEEE*, **15**(2), 1170–1177.

29 Chuang, S.T., Chen, T.Y., Chung, Y.C., Chen, R., & Lo, C.Y. (2015). Asymmetric fan-shape-electrode for high-angle-detection-accuracy tactile sensor. In: *Micro Electro Mechanical Systems (MEMS), 2015 28th IEEE International Conference on*, 18–22 January, 2015, Estoril, Portugal, pp. 740–743.

30 Lai, W.C., & Fang, W. (2015). Novel two-stage CMOS-MEMS capacitive-type tactilesensor with ER-fluid fill-in for sensitivity and sensing range enhancement. In: Solid-State Sensors, Actuators and Microsystems (TRANSDUCERS), 2015 Transducers-2015 18th International Conference on, 21–25 June, Alaska, pp. 1175–1178.

31 Hosono, M., Noda, K., Matsumoto, K., & Shimoyama, I. (2015). Dynamic performance analysis of a micro cantilever embedded in elastomer. *Journal of Micromechanics and Microengineering*, **25**(7), 075006.

32 Choi, E., Sul, O., Hwang, S., Cho, J., Chun, H., Kim, H., & Lee, S.B. (2014). Spatially digitized tactile pressure sensors with tunable sensitivity and sensing range. *Nanotechnology*, **25**(42), 425504.

33 Kim, U., Lee, D., Yoon, W., Hannaford, B., & Choi, H. (2015). Force sensor integrated surgical forceps for minimally invasive robotic surgery. *IEEE Transactions on Robotics*, **31**(5), 1214–1224.

34 Acer, M., Salerno, M., Agbeviade, K., & Paik, J. (2015). Development and characterization of silicone embedded distributed piezoelectric sensors for contact detection. *Smart Materials and Structures*, **24**(7), 075030.

35 Cutkosky, M., & Ulmen, J. (2014). Dynamic tactile sensing. In: R. Balasubramanian, V.J. Santos (eds.) *The Human Hand as an Inspiration for Robot Hand Development*, Springer Tracts in Advanced Robotics, vol. **95**, pp. 389–403.

36 Lee, J., Kwon, H., Seo, J., Shin, S., Koo, J.H., Pang, C., Son, S., Kim, J.H.; Jang, Y.H., Kim, D.E., & Lee, T. (2015). Conductive fiber-based ultrasensitive textile pressure sensor for wearable electronics. *Advanced Materials*, **27**(15), pp. 2409–2543.

37 Lang, W. (2011). From embedded sensors to sensorial materials the road to function scale integration. *Sensors and Actuators A: Physical*, **171**(1), pp. 3–11. From Embedded Sensors to Sensorial Materials Selected Papers from the E-MRS Spring Meeting 2010, Symposium A, Strasbourg (France), 2010.

38 Chen, X., Sakai, J., Yang, S., & Motojima, S. (2006). Biomimetic tactile sensors with fingerprint-type surface made of carbon microcoils/polysilicone. *Japanese Journal of Applied Physics*, **45**(37), L1019.

39 Takeshita, T., Harisaki, K., Ando, H., Higurashi, E., Nogami, H., & Sawada, R. (2016). Development and evaluation of a two-axial shearing force sensor consisting of an optical sensor chip and elastic gum frame. *Precision Engineering*, **45**, 136–142.

40 Sargeant, R., Liu, H., Seneviratne, L., & Althoefer, K. (2012). An optical multi-axial force/torque sensor for dexterous grasping and manipulation. In: *Multisensor Fusion and Integration for Intelligent Systems (MFI), 2012 IEEE Conference on*, 13–15 September, 2012, Hamburg, Germany, pp. 144–149.

41 Yeh, C.H. (2014). Image-assisted method for estimating local stiffness of soft tissues and calibration of bias due to aqueous humor effect. *Sensors and Actuators A: Physical*, **212**, 42–51.

42 Hu, Y., Katragadda, R., Tu, H., Zheng, Q., Li, Y., & Xu, Y. (2010). Bioinspired 3-D tactile sensor for minimally invasive surgery. *Journal of Microelectromechanical Systems*, **19**(6), 1400–1408.

43 Dosaev, M., Goryacheva, I., Martynenko, Y., Morozov, A., Antonov, F., Su, F.C., Yeh, C.H., & Ju, M.S. (2015). Application of video-assisted tactile sensor and finite element simulation for estimating Young's modulus of porcine liver. *Journal of Medical and Biological Engineering*, **35**(4), 510–516.

44 Kadmiry, B., Wong, C.K., & Lim, P.P. (2014). Vision-based approach for the localisation and manipulation of ground-based crop. *International Journal of Computer Applications in Technology*, **50**(1–2), 61–74.

45 Alvares, D. (2013). Development of nanoparticle film-based multi-axial tactile sensors for biomedical applications. *Sensors and Actuators A: Physical*, **196**, 38–47.

46 Yahud, S., Dokos, S., Morley, J., & Lovell, N. (2010). Experimental validation of a polyvinylidene fluoride sensing element in a tactile sensor. In: *Engineering in Medicine and Biology Society (EMBC), 2010 Annual International Conference of the IEEE*, 31 August–4 September, 2010, Buenos Aires, Argentina, pp. 5760–5763.

47 Yahud, S., Dokos, S., Morley, J., & Lovell, N. (2009). Experimental validation of a tactile sensor model for a robotic hand. In: *Engineering in Medicine and Biology Society, 2009. EMBC 2009. Annual International Conference of the IEEE*, 3–6 September, 2009, Minneapolis, MN, pp. 2300–2303.

48 Buchs, G., Maidenbaum, S., & Amedi, A. (2014). Obstacle identification and avoidance using the eyecane: a tactile sensory substitution device for blind individuals. In: M. Auvray, C. Duriez (eds.) *Haptics: Neuroscience, Devices, Modeling, and Applications*, Lecture Notes in Computer Science, vol. **8619**, pp. 96–103. Springer, Berlin/Heidelberg.

49 Herz, R., & Engen, T. (1996). Odor memory: review and analysis. *Psychonomic Bulletin and Review*, **3**(3), 300–313.

50 Ramic-Brkic, B., & Chalmers, A. (2010). Virtual smell: authentic smell diffusion in virtual environments. In: *Proceedings of the 7th International Conference on Computer Graphics, Virtual Reality, Visualisation and Interaction in Africa, AFRIGRAPH '10*, ACM, New York, 21–23 June, 2010, pp. 45–52.

51 Ghasemi-Varnamkhasti, M., Mohtasebi, S.S., Siadat, M., & Balasubramanian, S. (2009). Meat quality assessment by electronic nose (machine olfaction technology). *Sensors*, **9**(8), 6058.

52 Loutfi, A., Coradeschi, S., Kumar, G., Shankar, P., & Bosco, J. (2015). Electronic noses for food quality: a review. *Journal of Food Engineering*, **144**, 103–111.

53 Santos, J.P., Lozano, J., Aleixandre, M., Arroyo, T., Cabellos, J.M., Gil, M., & del Carmen Horrillo, M. (2010). Threshold detection of aromatic compounds in wine with an electronic nose and a human sensory panel. *Talanta*, **80**(5), 1899–1906.

54 Siadat, M., Losson, E., Ghasemi-Varnamkhasti, M., & Mohtasebi, S. (2014). Application of electronic nose to beer recognition using supervised artificial neural networks. In: *Control, Decision and Information Technologies (CoDIT), 2014 International Conference on*, 3–5 November, 2014, Metz, France, pp. 640–645.

55 Zheng, X.-Z., Lan, Y.-B., Zhu, J.-M., Westbrook, J., Hoffmann, W.C., & Lacey, R.E. (2009). Rapid identification of rice samples using an electronic nose. *Journal of Bionic Engineering*, **6**(3), 290–297.

56 Branca, A., Simonian, P., Ferrante, M., Novas, E., & Negri, R.M. (2003). Electronic nose based discrimination of a perfumery compound in a fragrance. *Sensors and Actuators B: Chemical*, **92**(12), 222–227.

57 oPhone duo (2014). http://www.engadget.com/2014/06/18/ophone-duo/ (accessed October 18, 2016).

58 Novich, S., & Eagleman, D. (2015). Using space and time to encode vibrotactile information: toward an estimate of the skins achievable throughput. *Experimental Brain Research*, **233**(10), 2777–2788.

59 Polshin, E., Rudnitskaya, A., Kirsanov, D., Legin, A., Saison, D., Delvaux, F., Delvaux, F.R., Nicolaï, B.M., & Lammertyn, J. (2010). Electronic tongue as a screening tool for rapid analysis of beer. *Talanta*, **81**(12), 88–94.

60 Rudnitskaya, A., Rocha, S.M., Legin, A., Pereira, V., & Marques, J.C. (2010). Evaluation of the feasibility of the electronic tongue as a rapid analytical tool for wine age prediction and quantification of the organic acids and phenolic compounds. the case-study of madeira wine. *Analytica Chimica Acta*, **662**(1), 82–89.

61 Cheung, I.W.Y., & Li-Chan, E.C.Y. (2014). Application of taste sensing system for characterisation of enzymatic hydrolysates from shrimp processing by-products. *Food Chemistry*, **145**, 1076–1085.

62 Medina-Plaza, C., Revilla, G., Muoz, R., Fernndez-Escudero, J.A., Barajas, E., Medrano, G., de Saja, J.A., & Rodriguez-Mendez, M.L. (2014). Electronic tongue formed by sensors and biosensors containing phthalocyanines as electron mediators: application to the analysis of red grapes. *Journal of Porphyrins and Phthalocyanines*, **18**(n01n02), 76–86.

63 Kortum, P. (2008). *HCI beyond the GUI: Design for Haptic, Speech, Olfactory and Other Nontraditional Interfaces*. Elsevier/Morgan Kaufmann, Amsterdam/Boston, pp. 291–306.

64 Derbyshire, D. (2009). The headset that will mimic all five senses and make the virtual world as convincing as real life. http://www.dailymail.co.uk/sciencetech/article-1159206/The-headset-mimic-senses-make-virtual-world-convincing-real-life.html (accessed November 4, 2016).

65 Ranasinghe, N., Cheok, A., Nakatsu, R., Yi-Luen Do, E. (2013). Simulating the sensation of taste for immersive experiences. In: *Proceedings of the 2013 ACM International Workshop on Immersive Media Experiences (ImmersiveMe '13)*. ACM, New York, pp. 29–34.

66 Arora, M. (2012). Basic principles of Sixth Sense technology. *VSRD-IJCSIT*, **2**(8), 687–693.

67 Rashmi, A., Bhatia, S., & Rani, G. (2012). Sensing the Sixth Sense technology. *International Journal of Information Technology and Knowledge Management*, **5**(1), 201–204.

68 Kohil, K. (2013). The Sixth Sense technology. http://prezi.com/n6dsa6yop3-0/the-sixth-sense-technology/ (accessed October 18, 2016).

69 Agarwal, D., Malhotra, L., & Jaiswal, A. (2014). Article: Sixth Sense technology: a variant for upcoming technologies. *International Journal of Computer Applications*, **102**(4), 20–25.

70 Kanel, K. (2014). The Sixth Sense technology. https://www.theseus.fi/bitstream/handle/10024/87120/final%20thesis_1_kedar.pdf?sequence=1 (accessed October 18, 2016).

71 VanSyckel, S., & Becker, C. (2014). A survey of proactive pervasive computing. In: Proceedings of the 2014 ACM International Joint Conference on Pervasive and Ubiquitous Computing: Adjunct Publication, UbiComp '14 Adjunct, ACM, New York, 13–17 September, 2014, Seattle, WA, pp. 421–430.

72 Billinghurst, M., Clark, A., & Lee, G. (2015). A survey of augmented reality. *Foundations and Trends in Human-Computer Interaction*, **8**(2–3), 73–272.

73 Cheng, H., Yang, L., & Liu, Z. (2015). A survey on 3-D hand gesture recognition. *IEEE Transactions on Circuits and Systems for Video Technology*, **26**(9), 1659–1673.

74 Bambach, S. (2015). A survey on recent advances of computer vision algorithms for egocentric video. *CoRR*, abs/1501.02825. https://arxiv.org/pdf/1501.02825v1.pdf.

75 Eagleman, D. (2015). The umwelt. http://www.eagleman.com/blog/umwelt (accessed October 19, 2016).

76 Wright, T., Margolis, A., & Ward, J. (2015). Using an auditory sensory substitution device to augment vision: evidence from eye movements. *Experimental Brain Research*, **233**(3), 851–860.

77 Proulx, M.J., Ptito, M., & Amedi, A. (2014). Multisensory integration, sensory substitution and visual rehabilitation. *Neuroscience and Biobehavioral Reviews*, **41**, 1–2.

78 Wright, T., & Ward, J. (2013). The evolution of a visual-to-auditory sensory substitution device using interactive genetic algorithms. *The Quarterly Journal of Experimental Psychology*, **66**(8), 1620–1638.

79 Hancock, P.A., Mercado, J.E., Merlo, J., & Erp, J.B.V. (2013). Improving target detection in visual search through the augmenting multi-sensory cues. *Ergonomics*, **56**(5), 729–738.

80 Salzer, Y., & Oron-Gilad, T. (2015). Evaluation of an "on-thigh" vibrotactile collision avoidance alerting component in a simulated flight mission. *IEEE Transactions on Human-Machine Systems*, **45**(2), 251–255.

81 Stronks, H.C., Parker, D.J., & Barnes, N.M. (2015). Tactile acuity determined with vibration motors for use in a sensory substitution device for the blind. *Investigative Ophthalmology and Visual Science*, **56**(7), 2920.

82 Crea, S., Cipriani, C., Donati, M., Carrozza, M., & Vitiello, N. (2015). Providing time-discrete gait information by wearable feedback apparatus for lower-limb amputees: usability and functional validation. *IEEE Transactions on Neural Systems and Rehabilitation Engineering*, **23**(2), 250–257.

83 Wu, S., Fan, R., Wottawa, C., Fowler, E., Bisley, J., Grundfest, W., & Culjat, M. (2010). Torso-based tactile feedback system for patients with balance disorders. In: *Haptics Symposium, 2010 IEEE*, 25–26 March, 2010, Massachusetts, pp. 359–362.

84 Gu, H., Kunze, K., Takatani, M., & Minamizawa, K. (2015). Towards performance feedback through tactile displays to improve learning archery. In: *Proceedings of the 2015 ACM International Joint Conference on Pervasive and Ubiquitous Computing and Proceedings of the 2015 ACM International Symposium on Wearable Computers, UbiComp '15*, ACM, 7–11 September, 2015, Osaka, Japan, pp. 141–144.

85 Hartcher-O'Brien, J., Auvray, M., & Hayward, V. (2015). Perception of distance-to-obstacle through time-delayed tactile feedback. In: *World Haptics Conference (WHC), 2015 IEEE*, 22–26 June, 2015, Chicago, IL, pp. 7–12.

86 Chebat, D.R., Maidenbaum, S., & Amedi, A. (2015). Navigation using sensory substitution in real and virtual mazes. *PLoS One*, **10**(6), e0126307.

87 Lee, B.C., Fung, A. (2015). Smartphone-based sensory augmentation technology for home-based balance training. *15th International Conference on Control, Automation and Systems (ICCAS 2015)*, 13–16 October, 2015, Busan, Korea.

88 Alfadhel, A., Khan, M.A., Cardoso, S., Leitao, D., & Kosel, J.. (2016). A magnetoresistive tactile sensor for harsh environment applications. *Sensors*, **16**(5), 650.

89 Melzer, M., Kaltenbrunner, M., Makarov, D., Karnaushenko, D., Karnaushenko, D., Sekitani, T., Someya, T., & Schmidt, O.G. (2015). Imperceptible magnetoelectronics. *Nature Communications*, **6**, 6080.

90 Warwick, K. (2014). A tour of some brain/neuronal–computer interfaces. In: G. Grübler, E. Hildt (eds.) *Brain-Computer-Interfaces in Their Ethical, Social and Cultural Contexts*, The International Library of Ethics, Law and Technology, vol. **12**, pp. 131–145. Springer, Dordrecht .

91 Warwick, K., Gasson, M., Hutt, B., Goodhew, I., Kyberd, P., Schulzrinne, H., & Wu, X. (2004). Thought communication and control: a first step using radiotelegraphy. *IEE Proceedings Communications*, **151**(3), 185–189.

92 Li, J., Li, Y., Zhang, M., Ma, W., & Ma, X. (2014). Cutaneous sensory nerve as a substitute for auditory nerve in solving deaf-mutes' hearing problem: an innovation in multi-channel-array skin-hearing technology. *Neural Regeneration Research*, **9**(16), 1532–1540.

93 Nghiem, B.T., Sando, I.C., Gillespie, R.B., McLaughlin, B.L., Gerling, G.J., Langhals, N.B., Urbanchek, M.G., & Cederna, P.S. (2015). Providing a sense of touch to prosthetic hands. *Plastic and Reconstructive Surgery*, **135**(6), 1652–1663.

94 Chuang, A.T., Margo, C.E., & Greenberg, P.B. (2014). Retinal implants: a systematic review. *British Journal of Ophthalmology*, **98**(7), 852–856.

95 Mannoor, M.S., Jiang, Z., James, T., Kong, Y.L., Malatesta, K.A., Soboyejo, W.O., Verma, N., Gracias, D.H., & McAlpine, M.C. (2013). 3D printed bionic ears. *Nano Letters*, **13**(6), 2634–2639.

96 Hansson, S. (2015). Ethical implications of sensory prostheses. In: J. Clausen, N. Levy (eds.) *Handbook of Neuroethics*, pp. 785–797. Springer, Dordrecht .

97 Lane, F., Nitsch, K., & Troyk, P. (2015). Participant perspectives from a cortical vision implant study: ethical and psychological implications. In: *Neural Engineering (NER), 2015 7th International IEEE/EMBS Conference on*, 22–24 April, 2015, Paris, France, pp. 264–267.

3

Advanced Reconfigurable 5G Architectures for Human Bond Communication

Enrico Del Re[1], Simone Morosi[1], Luca Simone Ronga[2], Lorenzo Mucchi[1], Sara Jayousi[1], and Federica Paganelli[2]

[1] Department of Information Engineering, University of Florence, Florence, Italy
[2] CNIT, Research Unit of Florence, Florence, Italy

3.1 Introduction

The improvement of the quality of our lives is the motivation that lies beneath the spreading of services and application whose objectives are the increase of comfort and security in our daily lives [1]. This goal is likely to push forward the concept of transmission of sensations by incorporating the five sensory features in the communications and allowing more expressive and holistic information exchange [2]. As highlighted in [3] and in the other chapters of this book, the possibility to integrate all the sensory features in the communications is still a research objective more than a development target.

Conversely, as for the communication of sensory information, the innovative and distinctive features of the upcoming 5G networks are a positive answer to the high level requirements, which human bond communication (HBC) applications demand for what concerns mobility, security, and quality of service (QoS): particularly, the broad set of possible HBC applications presents highly differentiated characteristics that nonetheless can be mapped in the extremely huge range of requirements that 5G systems afford [4, 5].

On the other hand, the introduction of the new paradigms of the software-defined networking (SDN) and network function virtualization (NFV) together with the evolution of the cloud computing and the integration of heterogeneous technologies will permit to define a flexible HBC architecture, which will be

Human Bond Communication: The Holy Grail of Holistic Communication and Immersive Experience, First Edition. Edited by Sudhir Dixit and Ramjee Prasad.

instantiated according to the specific requirements of each application and service: this specific architecture allows to adopt only the blocks, which are needed for the specific considered applications with a consistent reduction of used resources and a positive saving of required complexity.

To this aim in this chapter, we will show how the different HBC applications can be mapped in a general set of communication requirements and how the latter can be afforded in the HBC network architecture that will be proposed.

The chapter is organized as follows. Section 3.2 is devoted to the identification of high level requirements for the design of the network architecture for HBC services. Section 3.3 describes the proposed 5G network architecture based on SDN-NFV and fog/edge computing paradigms; moreover, the main enabling technological components for the provision of HBC services are highlighted. Section 3.4 presents some of the main security threats and possible solutions to be adopted. Finally, conclusions are drawn in Section 3.5.

3.2 HBC Communication Network Requirements

The objective of this section is to identify a set of requirements to be considered for the provisioning of future services, which will be based on the paradigm of HBC. Due to the large variety of application contexts that could be addressed, this is just a non-exhaustive list of high level requirements that enables a preliminary definition of the baseline system architecture that will guarantee the support for data gathering (from multiple and heterogeneous sources) and for data transmission over integrated communication systems. In the following the requirements have been grouped based on their purpose: general, network, QoS, and security.

3.2.1 General Requirements

In Table 3.1 the main general requirements for the design of the network architecture for HBC services are reported.

3.2.2 Network Requirements

In Table 3.2 some of the main network requirements for the exchange of data coming from the human five senses are reported.

3.2.3 Quality of Service Requirements

As already highlighted, the provision of advanced services, which involves the transmission of sensory information, relies on the availability of a network able to support very high data rates; however, in order to efficiently use the available resources and guarantee the required quality level, the adoption of QoS

Table 3.1 General requirements for HBC.

General requirement	Description
Integration of heterogeneous technologies	Legacy and advanced systems shall be seamlessly integrated in the future HBC baseline architecture for the user-transparent exploitation of the complementary capabilities of each system
Flexibility and scalability	The future HBC network architecture shall be able to self-adapt to the context conditions that may affect the provisioning of services (network resources availability, environmental changes, size of the users to be served etc.)
Availability and reliability	The user shall be able to access and manage the desired data anytime and anywhere, regardless of his location
Reconfigurability	The future HBC network architecture shall be highly reconfigurable in terms of network resources allocation, network nodes functions, network access technologies usage, and so on
Security and resilience	The future HBC network architecture shall provide highly robust security mechanisms for data transmission, distributed storage, and distributed access to personal data

Table 3.2 Network requirements for HBC.

Network requirement	Description
Data gathering	Multiple and heterogeneous sensors networks, including WBAN, shall be adopted for the collection and processing of human sensory information
Data access and transmission	Different technologies shall be selected or simultaneously used (aggregated bandwidth) for the efficient transmission of data, taking into account bandwidth and latency needs together with the specific context (network availability, user location, requested service)
Bandwidth	The network shall support very high data rates for the transmission of the human sensory information
	Very large bandwidth should be available for visual information transmission, while low bandwidth should be available for audio information transmission
	The required bandwidth for olfactory, gustatory, and tactile information has to be determined yet, but it depends on the density and types of sensory cells involved [7]
Latency	Latency shall be compliant with the specific application requirements (e.g., sensory information streaming is less susceptible to latency issues than real-time transmission)

Table 3.3 Quality of service requirements for HBC.

Network requirement	Description
Traffic management	The future HBC network architecture shall include traffic management mechanisms for the classification, prioritization, scheduling, and routing of the transmitted data
Classification	The data shall be classified based on different parameters (applications, priority etc.)
Priority	Specific priority shall be assigned to the different traffic flows to allow an accurate or personalized reconstruction of the human sensation on the receiver side
	High to very high priority is required for olfactory, gustatory, and tactile information transmission while medium priority for audiovisual information [7]
Scheduling	Scheduling techniques should be adopted for the proper reconstruction of the human sensation on the receiver side
Routing	Multipath routing techniques and data-centered routing mechanisms should be part of the future HBC baseline architecture

mechanisms shall be part of the network architecture for HBC services. Table 3.3 lists some of the main QoS requirements to be satisfied.

3.2.4 Security Requirements

One of the user's main needs is mobility. Acting in an open-access environment imposes to face the multiple threats characterizing the open wireless channel. Future HBC network architecture shall provide highly robust security mechanisms for data transmission, distributed storage, and distributed access to personal data.

In Table 3.4 the main security requirements are listed.

3.3 5G Architecture-Based SDN-NFV and Edge Computing for Human Bond Communication

The aim of this section is to describe the proposed network architecture, which will be able to efficiently support the provisioning of HBC services.

Taking into account all the previous identified high level requirements and considering the baseline HBC system architecture [2], new paradigms (SDN, NFV, and fog/edge computing) are introduced, and a highly flexible 5G architecture is presented.

Table 3.4 Security requirements for HBC.

	Security requirement	Description
Distributed data storage	Confidentiality	Data (especially person-related data) shall be kept confidential at a node or local server; data confidentiality should be resilient to device compromise attacks
	Integrity	Potential malicious modifications of data shall be detected all the time during the storage period and the user should be alerted
	Dependability	Data shall be readily retrievable even under node failure or malicious modifications
Distributed access to personal data	Data access control	A fine-grained access policy shall be defined to guarantee access to private information only by authorized parties (i.e., different access privileges for different users)
	Scalability	The distributed access control mechanism shall be scalable with the number of users in order to low management overhead of the access policies as well as low the computation and storage overhead
	Flexibility	The user shall be able to select the access point to connect for transmitting the personal data
	Accountability	Unauthorized actions on personal-related data shall be identified and the user should be informed about the prosecution of the abuse
	Revocability	The users/nodes privileges should be revoked if they are identified as compromised or malicious
	Non-repudiation	The origin of personal-related data cannot be denied later by the source that generated it. In other words, non-repudiation should protect a sender against the false assertion of the receiver that the message has not been received and a receiver against the false assertion of the sender that the message has been sent

3.3.1 Human Bond Communication: Baseline Architecture

To introduce the future HBC network architecture, it is worth starting to briefly analyze the baseline HBC architecture proposed in [2]. As depicted in Figure 3.1, it consists of three main entities:

1) **Senducers** (sense transducers). They perform sensory transduction of stimuli to electrical signals. In details, they translate the subjects in the human sensing domains, optimize the sense and perception, and communicate the data to the human bond sensorium (HBS) (described in the following).

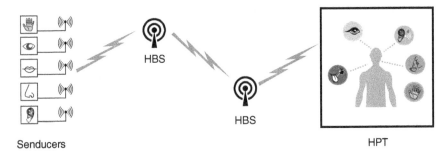

Senducers   HPT

Figure 3.1 HBC baseline system architecture [3].

2) **HBS.** It receives the sampled information (collected by the senducers), performs most of processing, and transmits the processed information. Therefore, HBS is responsible for data collection, processing, and transmission, including the implementation of security, QoS, and energy-saving mechanisms.

3) **Human perceivable transposer (HPT).** It is the last entity of the HBC architecture, and it is in charge of transposing the information received from HBS in human perceivable formats (i.e., into the sensory stimuli).

3.3.2 HBC Network Architecture

The recreation of the reality of the sensorial experience under the HBC paradigm relies on the definition of a robust and efficient communication system for the transport of sensed information. Particularly, the involved communication systems shall preserve the incisiveness and extensity features [2] satisfying the user needs in terms of quality and security.

Focusing on the compliance with the previously defined requirements in a mobile environment, the inclusion of 5G systems in the proposed HBC network architecture allows to guarantee high performance, in terms of throughput, latency, reliability, coverage, battery lifetime, and spectrum utilization. It is worth underlining that the 5G network, as the ubiquitous ultra-broadband network enabling the future Internet, will permit to redefine the network infrastructure as an extra-corporeal "nervous system" and enable "full immersive (3D) experience" enriched by "context information" and "anything or everything as a service (XaaS)" [4].

The envisaged 5G network evolution comes along with further significant innovations and changes in the networking and computing domains that let envisage a big revolution in the capability of ICT in supporting the human quality of life [10]. Two of them are:

1) A dynamic approach to networking enabled by the SDN [11] and NFV paradigms [12]
2) Evolution of cloud computing paradigms toward heterogeneous, federated, and distributed clouds as opposed to the current centralized model [13]

Focusing on the first one, SDN enables dynamic, programmable network connectivity by decoupling network control and data planes, while NFV allows for the on-demand instantiation and dynamic scaling of network functions on commodity hardware over the cloud [12]. The combination of these capabilities allows the creation of end-to-end virtual network paths and network services on demand.

As for the second one, cloud computing is extended to the edge of the network, paving the way to fog and mobile edge computing. Mobile edge computing [14] is enabled by the ability to run IT servers at network edge (e.g., at a base station site), thus allowing applications hosted on those servers to deliver low-latency and adaptive services by leveraging user proximity and local network information. Fog computing is a term coined by Cisco [15] referring to the extension of the cloud paradigm to the edge of the network (i.e., cloud close to the ground). The fog targets services and applications that demand widely distributed deployments and real-time interaction, and the Internet of things (IoT) is a main reference application scenario.

The envisaged 5G network architecture, enhanced also by the previously mentioned infrastructural innovations, is thus evolving toward a converged, cloud-integrated, ultra-broadband access network. In particular, the adoption of the virtualization paradigm in both the computing and networking domain envisages a landscape of heterogeneous service capabilities and resources pervasively distributed and interconnected and deeply integrated through the 5G network infrastructure. Thus, virtualization and "as-a-service" abstraction together pave the way toward a dynamic service ecosystem where dynamic data processing and delivery requirements can be promptly taken into account on a per user or a granular traffic flow basis [16]. Accordingly, service delivery paths can be dynamically established to serve an application data flow, which includes specified network and/or application processing functions to be traversed for addressing given application requirements.

Thus, such architectures can support the design and implementation requirements of the HBS component of the HBC architecture (Figure 3.1). Indeed, the resources needed for data processing, collection, and transmission tasks of the HBS could be flexibly and dynamically provisioned over this distributed NFV- and SDN-based 5G architecture. In fact, in HBC, the communication system is expected to handle the potentially different capacity and priority requirements of data flows originated by sensory transducers [7].

The comprehensive 5G vision described in METIS project [17] contains all the relevant components to satisfy the HBC requirements described previously: Figure 3.2 shows the main physical and logical layers of a 5G network as proposed by METIS project.

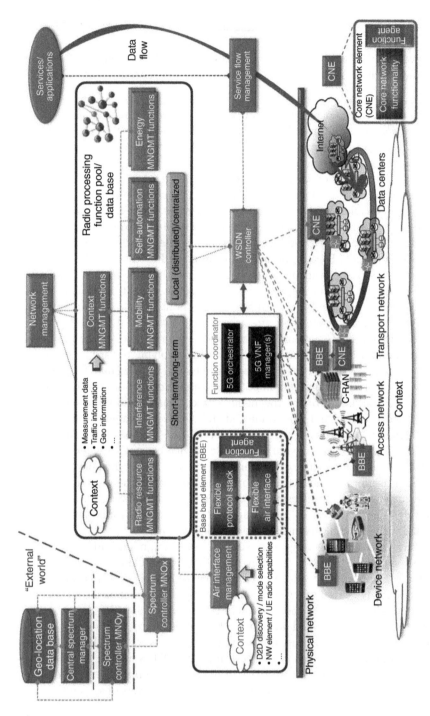

Figure 3.2 METIS 5G network architecture [17].

3.3.2.1 SDN-NFV for QoS Management

The key principle of SDN, that is, the decoupling of control and data planes, implies that the data plane is implemented by simple packet-forwarding devices, while the control plane is implemented by one or by a set of distributed controllers, which builds and maintains a logically centralized and abstract view of the network. This abstract view is offered through proper APIs to applications, which can thus access the control functions and program the network without delving into implementation details of the forwarding devices [18].

This approach thus promises to (i) radically simplify the development of network applications such as traffic engineering and network monitoring, (ii) allow granular flow classification and fine-grained and distributed QoS management policies enforcement (e.g., specific priority levels according to the type of sensory information), and (iii) dynamically steer traffic through chains of network functions for traffic optimization [9]. Such capabilities complement the ones offered by NFV. Indeed, thanks to the virtualization paradigm, network functions can be dynamically initiated, placed, migrated, and scaled to cope with general requirements (e.g., scalability, reconfigurability, and flexibility), as well specific network and security requirements (e.g., some network functions could be located at the network edge to minimize latency). Moreover, since the NFV paradigm leverages the use of general-purpose hardware, also application-level functions could be dynamically and selectively added to the chain of functions processing a specific traffic flow. Finally, end-to-end service life cycle management and SLA negotiation and assurance mechanisms across multiple domains can be implemented on top of SDN controller and applications and the NFV management and orchestration layer.

3.3.2.2 Fog/Edge Computing for HBC

The consolidated trend of the past decade of pushing computing, control, and storage into the cloud could not easily provide the requirements highlighted for HBC applications. Low latency, very large data volume from perceptive sensors, location awareness, and local utilization could promote strategies where computing and storage are moved toward the edge of the network moving the cloud down and close to devices (the "fog").

Fog computing is a key enabler for IoT and 5G mobile networks [19] since it owns properties that fit perfectly the challenging requirements for HBC. Synthetically,

- It provides security on resource constrained end point.
- Delivers low and predictable processing latency on nodes.
- Fog computing nodes (FCN) can host user-specific applications even in mobility.
- Enables device-to-device communications.
- It represents a performance boost for 5G radio access.
- Provides applications with awareness of the device position and context.
- It supports the NFV and separation of network control intelligence from radio network hardware.

Fog/edge computing HBC applications impact on both control and data planes. On the data plane the senducers data are cached and processed at the edge, reducing the storage requirements and making them available to local use via direct client-to-client communications. On the control plane the radio access network can be optimized by the distributed processing operated at end point; the 5G HetNets can be directly controlled by presence, amount, and priority of sensor data; local storage can deliver data during idle periods, reducing the impact on devices and network.

Fog-based radio access network (F-RAN) is the ideal choice for HBC. It provides an efficient and fast access to relevant data from sensors; it is available in large scale and follows the user node in mobility (Figure 3.3).

F-RAN architecture is structured with three layers: a cloud computing, a network access, and a terminal layer.

A fog-capable user equipment (F-UE) connect to fog access point (F-AP) where the distributed HBC application processes the incoming data and provides the required QoS signaling to network upper layers. Non-fog-capable UE can connect to remote radio head virtually operated by base band unit (BBU) in the cloud. Long-range UEs are served by high power nodes (HPN). Only fog-capable HBC nodes can fully exploit the features offered by fog architecture, but other nodes can still access the HBC service at a lower fidelity level.

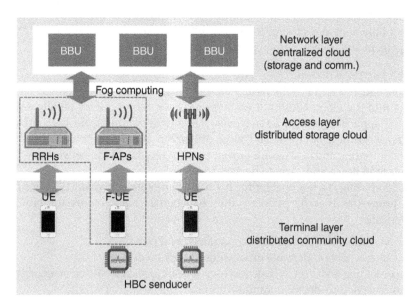

Figure 3.3 F-RAN HBC.

3.3.3 Main Building Blocks: HBC Enabling Technologies

3.3.3.1 WBAN for Data Gathering

As the sensorial information has to be gathered and conveyed once it has been sampled and interpreted by the senducers, it has to be transmitted to HBS where it will be transformed into a data set, by adding encryption and other security add-ons, and transmitted through a compatible medium.

The transmission of the acquired information will be performed by means of wireless local area networks (WLANs) or wireless body area networks (WBANs); in the former case the involved short-range communications will resort to the standards that are currently used for typical smart grid applications (IEEE 802.11, IEEE 802.15, ZigBee, etc.) or to the recently proposed D2D paradigm, which can be allowed also by the upcoming 5G cellular systems [21]. Conversely, the WBAN communications will be set and managed by wearable devices that will be capable of revealing the information, transforming it into a digital content and sending it through a communication channel. In this context, the different placements of the device on the body can be interpreted as the user's desire for a targeted application or the goal of affording "high fidelity" in the detection and extraction of the human stimulus. Therefore, this scenario can be seen as an evolution of the most advanced healthcare contexts where the WBAN has emerged as a new technology that allows the data of a patient's vital parameters and movements to be collected by small wearable or implantable sensors and communicated using wireless links. It is worth underlining that WBAN technology has a great potential in improving healthcare quality with a wide range of applications from ubiquitous health monitoring and computer-assisted rehabilitation to emergency medical response systems.

In the HBC vision, the data are not health-related but nonetheless they still retain the same importance, since they represent the senses and sensations of an individual.

3.3.3.2 Localization for Enhanced Services

In order to allow a complete provision and exploitation of the opportunities of the HBC applications, the adoption of accurate localization systems is of capital importance: in HBC scenarios the localization techniques will have to encompass the positioning not only of the user but also of the parts of the body as well as the relative position of the smart device on the body. This objective will enforce the current trends, which establish the accurate localization as a preliminary condition both for delivering new smart services (location-aware) and for outdoor and indoor navigation. Moreover, the capillary diffusion of sensors on bodies or objects will permit to have fine resolution and representation of the stimulus to be revealed or extracted and a fast and automatic sensor localization will permit the HBC devices to rapidly adapt to user's needs as well as interact autonomously with the body sensors.

As for the technologies that could be involved in the HBC systems, ultra-wide band (UWB) devices are naturally to be taken into account for accurate localization of smart devices and sensors on and in the body due to their inherent accuracy in short-range applications; besides, the UWB integration in multiple sensing devices can also be beneficial to overcome the limitation of individual technologies, in particular in harsh environments, [22] and to enhance the accuracy of other positioning technologies.

Since HBC systems will resort more to body-worn sensors than to external infrastructures such as the GNSS systems [8], the use of dead reckoning (DR) [23] techniques will be also enforced for the localization goals; in the framework of the HBC systems, the performance of the DR devices will be improved by the availability of low-cost micro-electromechanical systems (MEMS) sensors [24], which are already available on the market; moreover, these systems can benefit from the massive research activities, which have been recently focused on the definition of a more effective recognition of movement in the DR applications [23] in order to obtain a more precise position fix.

3.3.3.3 Content-Oriented Networking for HBC

Aiming at satisfying the quality level requirements of advanced application contexts, based on the distribution of a large amount of data, as HBC services are, besides the traditional networking concept, based on host-centric communications and addressing end points, the information-centric networking (ICN) approach should be considered [25].

ICN focuses on information objects, targeting to the design of a data-centric infrastructure that provides a mean to distribute contents in a scalable, cost-efficient, and secure manner. The main aim is to connect users to content.

In HBC context the transmission of sensation-related data can be performed through the adoption of new algorithms and protocols that efficiently locate and retrieve contents, satisfying the end user requirements in terms of quality of experience.

SDN and NFV both enable service-centric networking, but their paradigms will permit also to achieve the objective of ICN [25].

Moreover, in the framework of content-oriented networking techniques, user mobility and fast content retrieval, besides the adoption of multipath routing and content-aware resource mapping, could benefit from the use of mobile edge computing for dynamic content optimization [19]. The delivery of the sensory data can be enhanced taking into account the available location and network quality information.

3.4 HBC Security Issues and Potential Solutions

3.4.1 Main Security Threats

The HBC will happen to operate in open-access environments, which means that they can also accommodate attackers. The open wireless channel makes the data prone to being eavesdropped, modified, and jammed. Together with threats to stored data, other threats can come from the device compromise as well as from the network dynamic.

The sensor nodes worn by an HBC user are subjected to compromise, since they can be usually easily taken and opened or tampered. If data is directly encrypted and stored in a node along with its encryption key, the compromise of this node will lead to the disclosure of the data. In addition, local servers may not be trustworthy, since malicious people can either attack it remotely from Internet or simply go physically to the room where an HBC user is and wait for the chance to compromise the local server.

In addition, the HBC network dynamic is highly time variant, that is, nodes may join or leave the network frequently. Nodes may leave out due to low battery or due to a malicious attack. Other types of attacks may easily replace legitimate nodes with faked sensors and take away legitimate nodes with data inside. The personal-related data, if not well kept in more than one node, could be lost easily due to the network dynamics. Moreover, false data could be injected or treated as legitimate due to lack of authentication.

All these possible threats lead to stringent requirements when building the security of HBC.

Security threats can be located in each segment of the proposed network architecture, every time a wireless link connects two functional blocks. In addition, attacks can be moved to the data stored in the cloud. User's activities of daily living, while the user is using the device for HBC services, consist of data collection, processing, transmission, storing, and sharing. Major security threats are caused by the user's daily use and data sharing. HBC would be surely used for health applications, but probably the most used applications would be the social ones, which mainly mean sharing and exchanging of huge amount of personal data. This means that (i) the user will send personal data from many places potentially and (ii) the data flows into/from the network quite often every day. The expose to security breaks is obviously high. Moreover, the full mobility nature of HBC services implies that enough high quality connectivity must be ensured everywhere. The spectrum scarcity will force to move toward the cognitive radio to support such services in the next future.

As known, the cognitive paradigm of communication has been thought in order to use the scarcity of the spectrum with high efficiency. On the other hand, adding intelligence ("cognitivity") to the devices means more opportunities for malicious users [26]. The logic of primary and secondary users/networks can be embarrassed by a selfish user. For example, a primary selfish user can push energy in a channel in order to let it appear as occupied for secondary users. Or anyway, if the spectrum sensing result is modified maliciously, normal network activities will be disabled, and even the whole network traffic may be broken down [6].

HBC will require abundant use of the spectrum resource. Thus, cognitive radio approach will surely be a candidate to enable HBC in future wireless networks. Security issues are extremely important to be investigated and solved together with the development of a new communication service and system such as HBC. With HBC not only the classical personal data but also the very personal sensations of an individual are delivered wirelessly all over the world.

3.4.2 Possible Solutions

In order to support securely HBC services, which kind of solutions can be added to a communication system? This is not an easy task, but surely the future HBC system will require to add at least two main features to the traditional network security solutions: biometric encryption and physical-layer security. Biometric encryption would be used to protect the data from unauthorized access, while the physical-layer security would be used to make secure even the physical link where this information flows.

3.4.2.1 Biometric Encryption

Biometrics is the science of establishing the identity of an individual based on the physical, chemical, or behavioral attributes of the person [27]. The concept behind the biometric encryption is that the user is identified by some biological attribute that is unique for each individual. On the contrary, nowadays the user is authenticated/identified by something that he/she knows (password, code, etc.) or carries (SIM cards, chips, etc.). In other words, today it is not the user that is to be authenticated but an object that is carried by him/her. The other method is an alphanumerical password. It is well known that users tend to use weak passwords and to store the same password for all the sites requesting it. Using a biological parameter means you do not have to remember complicated password nor to carry with you any particular device or object when you want to access or send your data.

The sensation transductor can give an added value to biometrics. Not only biological attributes are unique in a human being (retina, fingerprint, DNA, etc.) but also sensations or, in particular, answers to stimuli. In this vision, the HBC can move the biometrics to "sensemetrics." The user's answer to a particular stimulus can be used for identification/authentication. For example, when the system needs to authenticate a user, it can send to the user a food taste or a smell and then evaluate the answer of the body.

The answer of the body to stimuli can also be used to encrypt the personal data of the user during the "journey in the Net." Sensations can be translated into a private digital key, which encrypts the data. Only the individual that has that body and can thus reproduce that answer to a specific stimulus can decrypt the data.

3.4.2.2 Physical-Layer Security

Today most used method for ensuring the security of data exchange in the Net is the so-called cryptography. Most commonly used cryptographic techniques, either asymmetric (based on public and private keys) or symmetric (based on a shared secret not known by others), are located at the upper layers of a network. But, actually, encryption does not protect from the undesired eavesdropping of the information but only from the interpretation of the data as meaningful text. In fact, the cryptographic techniques are based on the assumption that, statistically, the amount of time for performing a decrypting analysis is enormous. The amount of time to break a cipher text with a brute-force attack is related to the computational power of the attacker; in other words, the cryptography intrinsically assumes that the attacker has a limited amount of computational capability. Recent efforts of academia and industries to power up the amount of operations per second of the digital processors make this assumption weaker and weaker. When quantum computers will appear, how long should be a crypto key to ensure an enough long amount of time to break it?

Physical-layer security does not make any assumption on the computational power of the attacker [28]. In the past and in particular in wireless networks, the security has been added step by step to every layer of the ISO–OSI stack, except the physical one. The physical layer has always been devoted to ensure that the transmission went ahead well but not securely. Security at physical layer is historically mainly intended as the use of spread spectrum techniques (frequency hopping, direct sequence coding, etc.) in order to avoid the eaves-dropping. Moreover, scrambling the information data with a code does not assure a totally secure channel but just a long decryption time activity before getting the code by an unwanted listener, that is, the security is moved on the quantity of resources (hardware, time, etc.) that the unwanted listener must have in order to read the information.

A general definition of physical-layer security is set of mechanisms that exploit the properties of the physical layer to make an attacker's job harder.

The physical-layer security is based on the concept of perfect secrecy defined by Shannon [29] and developed later by Wyner [30]. Perfect secrecy is based on the fact that the randomness of the message is high enough to ensure that the eavesdropper cannot extract any useful information about the message by observing even for a long time the channel. A random source could be the noisy channel itself between the legitimate users. The studies over the physical-layer security has been increasing in the last decade by the scientific research community, and many different implementations have been proposed [31, 32], since together with the other layers, it can give a help in ensuring the security of a communication system. One thing is sure; the physical-layer security will be an added value for the wireless networks of the future like HBC, since it is the first step of the security stack, the first wall against malicious attacks.

3.5 Conclusions

The introduction of HBC paradigm in the provision of future services will enable a large variety of applications ranging from e-health to entertainment and from emergency to cognitive impairments management. The combined transmission of all the five senses information drives the user toward an augmented reality experience. Focusing on the communication technologies that will support these kinds of services, in this chapter, a novel HBC network architecture is proposed. The main requirements for the provision of HBC services are identified, and the baseline HBC system architecture reported in [2] is taken as a reference.

In order to define a high performance and highly reconfigurable architecture that efficiently meets the mobile user requirements for the HBC services fruition, a 5G-enabled network architecture based on the adoption of SDN-NFV and fog/edge computing paradigms is proposed. The main enabling technological components are described, focusing on WBAN, localization techniques, and content-oriented networking. To conclude, the main security threats for HBC services are identified and some possible solutions to be adopted are presented.

References

1 Del Re, E., Morosi, S., Ronga, L.S., Jayousi, S., Martinelli, A., "Flexible heterogeneous satellite-based architecture for enhanced quality of life applications," *IEEE Communications Magazine*, **53**(5), 186–193, 2015.

2 Prasad, R., "Human bond communication," *Wireless Personal Communications*, **87**(3), 619–627, 2016.

3 Del Re, E., Morosi, S., Mucchi, L., Ronga, L.S., Jayousi, S., "Future wireless systems for human bond communications," *Wireless Personal Communications*, **88**, 39, 2016.

4 Soldani, D., Pentikousis, K., Tafazolli, R., Franceschini, D., "5G networks: end-to-end architecture and infrastructure," *IEEE Communications Magazine*, **52**(11), 62–64, 2014.

5 5G—A Technology Vision, http://www.huawei.com/5gwhitepaper/ (accessed October 19, 2016).

6 Mucchi, L., Ronga, L.S., Del Re, E., "Physical layer cryptography and cognitive networks," *Wireless Personal Communications*, **58**(1), 95–109, 2011.

7 Wang, Y., Prasad, R., "Network neutrality impact on human bond communications," *Wireless Personal Communications*, **88**, 97, 2016.

8 Morosi, S., Martinelli, A., Del Re, E., "Peer-to-peer cooperation for GPS positioning," *International Journal of Satellite Communications and Networking*, 2016, doi: 10.1002/sat.1186.

9 Tomovic, S., Pejanovic-Djurisic, M., Radusinovic, I., "SDN based mobile networks: concepts and benefits," *Wireless Personal Communications*, **78**(3), 1629–1644, 2014.

10 Weldon, M.K., *The Future X Network: A Bell Labs Perspective*. CRC Press, Boca Raton, FL, 2016.

11 Kreutz, D., Ramos, F.M.V., Verissimo, P.E., Rothenberg, C.E., Azodolmolky, S., Uhlig, S., "Software-defined networking: a comprehensive survey," *Proceedings of the IEEE*, **103**(1), 14–76, 2015

12 Mijumbi, R., Serrat, J., Gorricho, J., Bouten, N., De Turck, F., Boutaba, R., "Network function virtualization: state-of-the-art and research challenges," *IEEE Communication Surveys and Tutorials*, **18**(1), 236–262, 2016.

13 European Commission. Horizon 2020, Work Programme 2016–2017. Information and Communication Technologies. European Commission Decision C (2015) 6776 of October 13, 2015. http://ec.europa.eu/research/participants/data/ref/h2020/wp/2016_2017/main/h2020-wp1617-leit-ict_en.pdf (accessed October 19, 2016).

14 ETSI, Mobile—Edge Computing Introductory Technical White Paper, September 2014. https://portal.etsi.org/Portals/0/TBpages/MEC/Docs/Mobile-edge_Computing_-_Introductory_Technical_White_Paper_V1%2018-09-14.pdf (accessed October 19, 2016).

15 Bonomi, F., Milito, R., Zhu, J., Addepalli, S., "Fog computing and its role in the internet of things," Proceedings of the First Edition of the MCC Workshop on Mobile Cloud Computing (MCC'12), ACM, New York, NY, August 17, 2012, pp. 13–16.

16 Martini, B., Paganelli, F., Cappanera, P., Turchi, S., Castoldi, P., "Latency-aware composition of Virtual Functions in 5G," 2015 First IEEE Conference on Network Softwarization (NetSoft), IEEE, London, April 17, 2015, pp. 1–6.

17 Monserrat, J.F., Tullberg, H., Mange, G., Zimmermann, G., "METIS research advances towards the 5G mobile and wireless system definition," *EURASIP Journal on Wireless Communications and Networking*, 2015(1), 1–16, **2015**.

18 Stallings, W., *Foundations of Modern Networking*, Pearson Education, Indianapolis, 2015.

19 Borcoci, E., "Fog-computing versus SDN/NFV and cloud computing in 5G," DataSys 2016 Conference, Valencia, Spain, May 22, 2016.

20 Peng, M., Yan, S., Zhang, K., Wang C., "Fog computing based radio access networks: issues and challenges," *IEEE Network*, **30**(4), 46–53, 2016.

21 Malandrino, F., Casetti, F., Chiasserini, C.F., "Toward D2D-enhanced heterogeneous networks," *IEEE Communications Magazine*, **52**(11), 94–101, 2014.

22 Conti, A., Dardari, D., Guerra, M., Mucchi, L., Win, M.Z., "Experimental characterization of diversity navigation," *IEEE Systems Journal*, **8**(1), 115–124, 2014

23 Martinelli, A., Morosi, S., Del Re, E., "Daily living movement recognition for pedestrian dead reckoning applications," *Mobile Information Systems*, 2016, 7128201, **2016**.

24 Yazdi, N., Ayazi, F., Najafi, K., "Micromachined inertial sensors," *Proceedings of the IEEE*, **86**, 16401659, 1998.

25 TalebiFard, P., Ravindran, R., Chakraborti, A., Pan, J., Mercian, A., Wang, G., Leung, V.C.M., "An Information Centric Networking approach towards contextualized edge service," 2015 12th Annual IEEE Consumer Communications and Networking Conference (CCNC), Las Vegas, NV, IEEE, Piscataway, NJ, January 9–12, 2015, pp. 250–255.

26 Mucchi, L., Carpini, A., "Aggregate interference in ISM band: WBANs need cognitivity?," 2014 Ninth International Conference on Cognitive Radio Oriented Wireless Networks and Communications (CROWNCOM), Oulu, Finland, June 2–4, 2014, pp. 247–253.

27 Jain, A.K., Flynn, P., Ross, A.A., *Handbook of Biometrics*. Springer-Verlag, New York, 2007.

28 Mucchi, L., Ronga, L.S., Cipriani, L., "A new modulation for intrinsically secure radio channel in wireless systems," Special issue "Information security and data protection in future generation communication and networking," *Wireless Personal Communications*, **51**(1), 67–80, 2009.

29 Shannon, C.E., Author, T., "Communication theory of secrecy systems," *Bell System Technical Journal*, **29**, 166, 656–715, 1949.

30 Wyner, A.D., Author, T., "The wire-tap channel," *Bell System Technical Journal*, **54**, 1355–1387, 1975.

31 Zhou, X., Song, L., Zhang, Y., *Physical Layer Security in Wireless Communications*, CRC Press, Boca Raton, FL, 2013.

32 Mukherjee, A., Fakoorian, S.A.A., Huang, J., Swindlehurst, A.L., "Principles of physical layer security in multiuser wireless networks: a survey," *IEEE Communication Surveys and Tutorials*, **16**(3), 1550–1573, 2014.

4

Data Mining of the Human Being

Mauro De Sanctis and Pierpaolo Loreti

Interdepartmental Center for Teleinfrastructures (I-CTIF), University of Rome "Tor Vergata", Rome, Italy

4.1 Introduction

The concept of data mining can be dated back to 1990. Data mining is also known as knowledge discovery in databases (KDD) or less frequently as knowledge discovery and data mining. While a commonly accepted definition is not available, we can define data mining as the process of analyzing data from different perspectives and extracting hidden information, identifying patterns, or relationships among the data. In this chapter we focus on data mining of the human being, hence, the data is any fact, number, or text regarding a human being.

Human life is organized in a hierarchical manner where lower levels of organization are progressively integrated with increasing complexity to build up higher levels. Organization levels, starting from the lower levels, include atoms, molecules (such as an amino acid or a nucleotide), cells (e.g., a neuron or a skin cell), tissues (e.g., nervous tissue or epithelial tissue), organs (the heart or the brain), organ systems (e.g., the cardiovascular or the digestive systems), organism (i.e., the human), and populations of humans together with their social interaction. Therefore, in this chapter, data describe the human being at any level, from atoms to cells, to organs, and to social level.

The composition of the human body can be described from the point of view of either mass composition or atomic composition. The elements present in the total body of the so-called "standard man" expressed by percentage of mass are the following: oxygen (65%), carbon (18%), hydrogen (10%), nitrogen (3%), calcium (1.5%), phosphorus (1%), sulfur (0.25%), potassium (0.2%), sodium (0.15%), chlorine (0.15%), magnesium (0.05%), iron (0.006%), and zinc (0.003%). We are not including traces of chemical elements below 0.003% of the total body mass [1].

Human Bond Communication: The Holy Grail of Holistic Communication and Immersive Experience, First Edition. Edited by Sudhir Dixit and Ramjee Prasad.
© 2017 John Wiley & Sons, Inc. Published 2017 by John Wiley & Sons, Inc.

Chemical elements of the human body have the capability to perform certain structural functions or to provide specific reactivities. In fact, carbon forms multiple covalent bonds with other carbon atoms as well as with other elements such as nitrogen, hydrogen, oxygen, or sulfur. This feature allows the construction of long carbon chains and rings with the presence of reactive functional groups containing nitrogen, oxygen, and sulfur as in proteins, nucleic acids, lipids, and carbohydrates.

The cell is the fundamental unit of life consisting of cytoplasm enclosed within a membrane, which contains many biomolecules. A cell is the smallest unit of life that can replicate independently; humans contain more than 10 trillion cells.

Tissues are made up of cells that are similar in structure and function; cells of a tissue work together to perform a specific activity. The human body is composed of four basic types of tissues: epithelium, connective, muscular, and nervous. These tissues vary in their composition and function.

An organ is made from a group of different tissues, which all work together to do a particular job.

Although each organ has its specific function, organs also function together in groups, called organ systems. Medical doctors categorize disorders and their own medical specialties according to organ systems.

The complexity of the human being makes data mining an important tool to gain new knowledge about human life. Here we refer to data mining as the process of analyzing a set of data (also dataset) from different perspectives with the aim to extract hidden information and to identify patterns or relationships among the data. A dataset can be organized as an $M \times N$ data matrix $\mathbf{X}$, with each row representing an *instance* (also data point, record, or example) and each column containing the value of a *feature* (also attribute or variable) for each instance. Therefore, the data matrix $\mathbf{X}$ contains M instances $\mathbf{x}_i = [x_{i,1}, x_{i,2}, ..., x_{i,N}]$ for $i = 1, 2, ..., M$.

There are two different types of features: (i) a numerical feature that has a real-valued or integer-valued domain and (ii) a categorical feature (also symbolic feature) that has a set-valued domain.

The data mining process includes the data preprocessing phase and the data processing phase.

The data preprocessing phase may include one or more of the following tasks:

- Data management and transformation of format
- Discretization
- Outlier and missing value management
- Dimensionality reduction
- Data normalization
- Feature extraction
- Feature selection

The data processing phase involves learning methods for descriptive and predictive data analysis, visualization, and online updating. In particular, descriptive methods (also one-step unsupervised learning methods) reveal properties of data, while predictive methods (also two steps supervised learning methods) use historical data (training data) to predict future data. In the following, we summarize both descriptive and predictive methods.

4.1.1 Descriptive Methods

- *Association rule mining* is the process of identifying the rules of dependence between two or more features. It can be considered as a method to discover local or global association rules between feature values or as a method to discover frequent joint values of features in the dataset (frequent itemsets). Association rule mining algorithms include Apriori and frequent pattern growth (FP-Growth).
- *Clustering (or segmentation)* is a method to divide the instances of a dataset into *G* groups (clusters) according to similarities within each group (homogeneity). Well-known clustering algorithms include k-means, density-based spatial clustering of applications with noise (DBSCAN), ordering points to identify the clustering structure (OPTICS), and balanced iterative reducing and clustering using hierarchies (BIRCH).
- *Sequential pattern mining* discovers frequent ordered list of events or subsequences in a dataset. Mining sequential patterns can be viewed as a subset of the problem of mining association rules between dataset elements, constrained by the temporal aspects of the data. Popular sequential pattern mining algorithms include prefix-projected sequential pattern mining (PrefixSpan), generalized sequential pattern (GSP), and sequential pattern discovery using equivalent class (SPADE).

4.1.2 Predictive Methods

- *Classification*: In a classification problem, the dataset contains one or more categorical (or discrete) features called class features. A classification model uses historical data to predict the class feature of a new instance on the basis of the values of the instance. In particular, classification algorithms assign a new instance to a class using its similarity with historical instances belonging to that class. The most popular classification algorithms include k-nearest neighbors (k-NN), support vector machines (SVM), Bayes, naive Bayes, ID3, C4.5, artificial neural networks (ANN), linear discriminant analysis (LDA), and decision trees.
- *Regression*: It refers to the problem of attempting to build a regression function that allows to determine the relationship between the dependent variable (target feature) and one or more independent variables. Regression models are similar to classification models. However, regression deals with

continuous numerical target features, whereas classification deals with categorical or discrete numerical target features. Popular regression algorithms include linear regression and logistic regression.

- *Anomaly detection*: It is the task of identifying unusual patterns that do not conform to the expected behavior of the data, or it predicts whether an instance is typical for a given distribution or not. It can be addressed as (i) a classification method where the class feature only contains two possible values, (ii) a clustering method where new instances not belonging to any cluster or belonging to clusters of low density are considered as outliers, and (iii) a discovery of deviations from association rules and frequent itemsets.

For each organization level of the human being, a data mining method can be carried out using one out of many different algorithms.

4.2 Data Mining in Molecular Biology

4.2.1 Genomics

A revolutionary advance in biological research was achieved by the concepts of molecular biology, which links information about genetic traits to deoxyribonucleic acid (DNA).

The human genome is the complete set of nucleic acid sequence for humans, encoded as DNA within the 23 pairs of chromosomes in cell nuclei and in a small DNA molecule found within individual mitochondria. In humans, a copy of the entire genome is contained in all cells that have a nucleus.

The length of the human genome sequence is about 3 billion nucleotides (base pairs). Nucleotides are organic molecules that serve as the subunits of DNA and ribonucleic acid (RNA). A gene is a region of the DNA, which is made up of nucleotides and is the molecular unit of heredity. Gene expression is the process by which the nucleotide sequence of a gene is used to drive protein synthesis and produce the structures of the cell. RNA is responsible for coding, decoding, regulation, and expression of genes. Messenger RNA (mRNA) is a large family of RNA molecules that convey genetic information.

The Human Genome Project (HGP) was the international collaborative research program whose goal was the complete mapping and understanding of the human genome.

In large-scale genomic studies, data is analyzed with two different objectives in mind. The first objective is to identify genetic variants that may play a role (association rule mining problem), either alone or in concert, in disease development or progression. The second objective is to classify or predict disease development or progression based on genetic variants, again either alone or in combination (classification problem).

Gene expression data includes the expression level of a large number of genes of an organism for a number of different experimental samples (conditions). The samples may correspond to different time points or different environmental conditions. In other cases, the samples may have come from different organs, from cancerous or healthy tissues or even from different individuals [2].

Usually, gene expression data is arranged in a data matrix, where each gene corresponds to one row and each condition to one column. Each element of this matrix represents the expression level of a gene under a specific condition and is represented by a real number, which is usually the logarithm of the relative abundance of the mRNA of the gene under the specific condition.

Objectives of clustering algorithms for gene expression data include (i) grouping genes by their expressions over conditions/samples, (ii) grouping conditions/samples based on the expression of genes, and (iii) finding subgroups of genes and conditions/samples such that the identified genes share similar expression patterns over a specified subset conditions/samples.

Association rule mining has been also applied in Ref. [3] to the analysis of gene expression data, in order to reveal biologically relevant associations among different genes or between environmental effects and gene expression.

4.2.2 Proteomics

Protein is a biological macromolecule composed of one or more chains of amino acids. Proteins differ from one another primarily in their sequence of amino acids, which usually results in a specific three-dimensional (3-D) structure that determines the protein activity.

Proteins perform a vast array of functions, defending the body from antigens (antibodies), facilitating biochemical reactions (enzymes), moving molecules from one place to another around the body (transport proteins), stimulating certain bodily activities (hormones), and so on.

The proteome is the entire set of proteins produced or modified by an organism. Scientists in the area of proteomics aim to classify and characterize the proteins, study their interactions with other proteins, and identify their functional roles. Many functional aspects of proteins are determined mainly by local sequence characteristics and do not depend critically on the full 3-D structure.

Protein analysis can be carried out through the matrix-assisted laser desorption/ionization (MALDI), which is an ionization method used in mass spectrometry for the analysis of protein mixtures. Surface-enhanced laser desorption/ionization (SELDI) is a variation of MALDI, which is also used for protein analysis.

The applicability of a genetic algorithm and a classification method based on k-NN to SELDI proteomics data analysis was demonstrated in Ref. [4].

In this work it is shown that a k-NN classification algorithm is effective in finding a few ions capable of reliable discrimination between cancer and unaffected serum specimens in a SELDI dataset.

In Ref. [5], several classification algorithms have been tested using blood plasma MALDI data to identify significant and specific differences between heart failure patients and healthy individuals. The SVM algorithm gave the most promising results achieving an accuracy of 95% using 18 selected features (biomarkers).

Association rule mining algorithms are applied in Ref. [6] to discover rules that relate protein properties (e.g., functional annotation sequence motifs) and in Ref. [7] to protein–protein interactions.

4.2.3 Molecule Mining

Biologists aim to find new drug candidates analyzing hundreds of thousands of molecules. Designing new medical drugs for a particular disease requires the analysis of the molecules that have an activity for that disease with the objective to discover substructures (called fragments) that are correlated to the activity of these molecules [8]. The relationship between the 3-D graph representation of a molecule or other descriptors and its biological activity is called structure–activity relationship (SAR). There are two types of SAR methods:

1) Quantitative SAR methods aim to find a correlation between the descriptors of a molecule and its activity. Descriptor vectors are computed for different molecules and arranged in a dataset. Then, different data mining techniques are applied to the dataset. Many different classification algorithms are used in quantitative SAR where the class attribute is the response variable.
2) Qualitative SAR methods aim to find some common substructures in the structure of molecules. There are many approaches related to qualitative SAR in literature, which are based on frequent pattern mining methods. The underlying concept is that similar compounds have similar activity or dissimilar compounds have dissimilar activity.

The fragments that demonstrate high correlation to the activity for a disease are then used to design new medicines for that specific disease [9].

4.3 Data Mining in Cytology and Histology

The human body has many different kinds of cells. Though they might look different under a microscope, most cells have chemical and structural features in common. In humans, there are about 200 different types of cells. All cells have a membrane, which is the outer layer that holds the cell together. The membrane lets nutrients pass into the cell and waste products pass out.

The cell have a nucleus containing the DNA coding for proteins and all the apparatus necessary for maintaining life, that is, enzymes, multiprotein complexes, and so on. Cytoplasm is the viscous aqueous medium between cell membrane and nucleus.

Data mining algorithms can be applied to both cell images and to cell features to automatically recognize cell type.

A first example of cell type dataset is used in Ref. [10]. In this work, features are computed from a digitized image of a fine needle aspirate of a breast mass. These features describe the characteristics of the cell nuclei present in the image. A class label describes the diagnosis, M = malignant, B = benign. The proposed classification algorithms achieve a very high accuracy in the recognition of malignant or benign cells.

The data mining method described in Ref. [11] provides a prescreening for automated cytological analysis based on nanocytology using partial wave spectroscopy (PWS), which enables quantification of the statistical properties of cell structures at the nanoscale. The method was tested on buccal cytology but can easily be extended to other types of cytological samples. This automated technique may be a valuable method of cell selection with particular relevance to identify patients harboring premalignant tumors.

Authors of Ref. [12] present the application of a genetic algorithm and an SVM classification method to the recognition of blood cells based on using the image of the bone marrow aspirate. The main task of the genetic algorithm is the selection of the features used by the SVM in the final recognition and classification of cells.

4.4 Medical Data Mining

The human body is made up of several organ systems that all work together as a unit to make sure the body keeps functioning. In fact, each organ system depends on the others, either directly or indirectly, to keep the body functioning effectively. The organ systems of the human body are as follows:

- Circulatory system, which is responsible to move the blood around the body.
- Digestive system (or gastrointestinal system), which is responsible to eat and digest food.
- Endocrine system, which allows to regulate the metabolism and plays a part in growth, development, tissue function, and mood.
- Integumentary system, which includes everything covering the outside of an animal's body, that is, skin, hair, nails, and sweat glands.
- Muscular system, which is responsible to produce movements.
- Reproductive system, which is responsible for sexual reproduction.
- Respiratory system, which consists of specific organs and structures used for the process of respiration.

- Skeletal system, which performs vital functions such as support, movement, protection, blood cell production, calcium storage, and endocrine regulation.
- Urinary system, which produces, stores, and eliminates urine, the fluid waste excreted by the kidneys.
- Immune system, which can detect and identify many different kinds of disease to protect the human body from foreign bodies.
- Nervous system, which is responsible to collect, transfer, and process information with the brain, spinal cord, and peripheral nervous system. It allows humans to respond to what is around them.

Medical data managed by doctors for evaluating diagnosis and therapy has a variety of formats: images in the form of X-rays or scans, textual information to describe details of diseases, medical histories, psychology reports, medical articles, or various biosignals like electrocardiogram (ECG or EKG), electroencephalogram (EEG), and so on [13–15]. The general current model of healthcare dictates that medical practitioners and other actors operating in the healthcare system keep their own individual records of patients. Digital health records are classified as follows:

- Electronic medical records (EMRs): They are digital versions of the paper charts in clinician offices. EMRs contain notes and information collected by clinicians in their offices and are mostly used for diagnosis and treatment. EMRs are more valuable than paper records because they enable clinicians to track data over time, identify patients for preventive visits and screenings, monitor patients, and improve healthcare quality.
- Electronic health records (EHRs): They are built to go beyond standard clinical data collected in a clinician's office and are inclusive of a broader view of a patient's care. EHRs contain information from a set of clinicians involved in a patient's care (e.g., a hospital or a municipality).
- Personal health records (PHRs): They can contain the same types of information as EMRs and EHRs (e.g., diagnoses, medications, immunizations, family medical histories, and provider contact information), but they are designed to be set up, accessed, and managed by patients only. Patients can use PHRs to maintain and manage their health information in a private, secure, and confidential environment. PHRs can include information from a variety of sources including home monitoring devices and patients themselves.

The challenges for data mining in health records overlap in large part with the well-known challenges for big data analytics. They are the following:

- Use of unstructured datasets: Usually digital prescriptions and medical reports as entered by the physicians are narrative and unstructured. This requires the use of text mining tools, which are sources of errors.

- Statistical validity: Typical health records contain few data points (for a single event and a single individual) and many features. This means that the statistical modeling through data mining methods becomes very challenging and the validity of the results is limited.
- Complexity and speed of data mining algorithms: When we deal with high dimensional data, that is, a large number of records and/or features, the complexity of the data mining algorithms has an impact on the duration of the data analysis.
- Dealing with outliers: An outlier is a record that is unlike most others in the dataset because of many possible reasons. In large datasets it is not easy to find outliers, understand if they come from errors in the data collection process, and eventually decide to discard them.
- Displaying meaningful results: The display of results in a meaningful format for the analyst is challenging when a large number of features is considered, as it is for typical digital health records.

A growing trend is noted in the use of classification methods toward cancer prediction integrating features such as family history, age, diet, weight, high-risk habits, and exposure to environmental carcinogens [16–20].

In medicine, decision support systems (DSS) refer to a class of computer-based systems that aids the doctors during the process of decision-making.

A fuzzy rule-based DSS is presented in Ref. [21] for the diagnosis of coronary artery disease. The dataset used for the DSS generation and evaluation consists of 199 subjects, each one characterized by 19 features, including demographic and history data, as well as laboratory examinations.

A feature selection scheme via supervised model construction was applied in Ref. [22] to identify the key factors affecting blood glucose control among type 2 diabetic patients. It was shown that several classifiers are able to achieve their best performance when the top 15 features are selected. In this research, it was found that the major discriminative features are age, diagnosis duration, insulin treatment, random blood glucose, and diet treatment. While age and duration of diagnosis are variables that cannot be controlled, blood glucose levels can be positively influenced by the management of the disease.

4.5 Opinion Mining

What other people think has always been an important social aspect for most of us during the decision-making process [23, 24]. Internet and the Web have now, among other things, made it possible to find out about the opinions and experiences of people we have never heard of.

Opinion mining, sentiment analysis, and subjectivity analysis are interrelated areas of research that use various techniques taken from natural language processing (NLP), information retrieval (IR), and structured and unstructured data mining. These terms enclose sentiment, opinions, emotions, evaluations, beliefs, and speculations. Opinion mining is widely applied to reviews and social media for a variety of applications, ranging from marketing to customer service.

Pang et al. [25] classified movie reviews into two classes, positive and negative. It was shown that using unigrams (a bag of individual words) as features in classification performed well with either naive Bayesian or SVM.

Moraes et al. present an empirical comparison between SVM and ANN regarding document-level sentiment analysis [26]. They adopted a standard evaluation context with popular supervised methods for feature selection and weighting in a traditional bag-of-words model. Except for some unbalanced data contexts, the experiments indicated that ANN produce superior or at least comparable results to SVMs.

Twitter messages are increasingly used to determine consumer sentiment toward a brand. Ghiassi et al. introduced an approach to supervised feature reduction using n-grams and statistical analysis to develop a Twitter-specific lexicon for sentiment analysis [27]. They augmented this reduced Twitter-specific lexicon with brand-specific terms for brand-related tweets. Results show that the reduced lexicon set, while significantly smaller (only 187 features), reduces modeling complexity, maintains a high degree of coverage over the Twitter corpus, and yields improved sentiment classification accuracy.

4.6 Conclusions

In this chapter we provided an overview of data mining applications at different levels of organization of the human being, from molecular level to social level. However, given the vastness of the subject matter, this chapter can only aspire to provide a starting point for further considerations and research.

Today, most of the studies involving gene or protein expression data is performed exclusively by biologists or biomedical scientists. However, the importance of including data scientists in such studies is indisputable [28]. An effective interdisciplinary approach to the research is what will make the difference in future studies.

The interdependence between different levels of organizations, for example, molecular (genetic), organ systems, social behavior, and interaction (family history, work, lifestyle), should be considered as a more complete approach to study causes and effects on the human being.

Furthermore, since we are dealing with the human being, privacy and ethical issues should be always considered in our future works.

References

1 Spiers, F. W., *"Radioisotopes in the Human Body: Physical and Biological Aspects,"* Academic Press, New York, 1968.

2 Madeira, S. C., Oliveira, A. L., "Biclustering algorithms for biological data analysis: a survey," *IEEE/ACM Transactions on Computational Biology and Bioinformatics*, vol. **1**, no. 1, pp. 24–25, January 2004.

3 Creighton, C., Hanash, S., "Mining gene expression databases for association rules," *Bioinformatics*, vol. **19**, no. 1, pp. 79–86, January 2003.

4 Li, L., Umbach, D. M., Terry, P., Taylor, J. A., "Application of the GA/KNN method to SELDI proteomics data," *Bioinformatics*, vol. **20**, no. 10, pp. 1638–1640, July 2004.

5 Willingale, R., Jones, D. J., Lamb, J. H., Quinn, P., Farmer, P. B., Ng, L. L., "Searching for biomarkers of heart failure in the mass spectra of blood plasma," *Proteomics*, vol. **6**, pp. 5903–5914, November 2006.

6 Kotlyar, M., Jurisica, I., "Predicting protein-protein interactions by association mining," *Information Systems Frontiers*, vol. **8**, no. 1, pp. 37–47, February 2006.

7 Oyama, T., Kitano, K., Satou, K., Ito, T., "Extraction of knowledge on protein-protein interaction by association rule discovery," *Bioinformatics*, vol. **18**, pp. 705–714, May 2002.

8 Yılmaz, B., Göktürk, M., "Interactive data mining for molecular graphs," *Journal of Automated Methods and Management in Chemistry*, vol. **2009**, no. 2, pp. 1–12, October 2009.

9 Holzinger, A., Dehmer, M., Jurisica, I., "Knowledge discovery and interactive data mining in bioinformatics—state-of-the-art, future challenges and research directions," *BMC Bioinformatics*, vol. **15**, Suppl. 6, pp. 1–9. May 2014.

10 Wolberg, W. H., Street, W. N., Heisey, D. M., Mangasarian, O. L., "Computer-derived nuclear features distinguish malignant from benign breast cytology," *Human Pathology*, vol. **26**, no. 7, pp. 792–796, July 1995.

11 Miao, Q., Derbas, J., Fid, A., Subramanian, H., Backman, V., "Automated cell selection using support vector machine for application to spectral nanocytology," *BioMed Research International*, vol. 2016, no. 3, pp. 1–10, January 2016.

12 Osowski, S., Siroic, R., Markiewicz, T., Siwek, K., "Application of support vector machine and genetic algorithm for improved blood cell recognition," *IEEE Transactions on Instrumentation and Measurement*, vol. **58**, no. 7, pp. 2159–2168, August 2009.

13 Aragüés, A., Escayola, J., Martínez, I., del Valle, P., Muñoz, P., Trigo, J. D., García, J., "Trends and challenges of the emerging technologies toward interoperability and standardization in e-health communications," *IEEE Communications Magazine*, vol. **49**, no. 11, pp. 182–188, November 2011.

14 Bousquet, J. et al., "Systems medicine and integrated care to combat chronic noncommunicable diseases," *Genome Medicine*, vol. **3**, pp. 43, July 2011.

15 Bellazzi, R., Zupan, B., "Predictive data mining in clinical medicine: current issues and guidelines," *International Journal of Medical Informatics*, vol. **77**, no. 2, pp. 81–97, February 2008.

16 Richards, G., Rayward-Smith, V., Sonksen, P., "Data mining for indicators of early mortality in a database of clinical records," *Artificial Intelligence in Medicine*, vol. **22**, no. 3, pp. 215–231, June 2001.

17 Bach, P. B., et al., "Variations in lung cancer risk among smokers," *Journal of the National Cancer Institute*, vol. **95**, no. 6, pp. 470–478, March 2003.

18 Domchek, S. M., Eisen, A., Calzone, K., Stopfer, J., Blackwood, A., Weber, B. L., "Application of breast cancer risk prediction models in clinical practice," *Journal of Clinical Oncology*, vol. **21**, no. 4, pp. 593–601, February 2003.

19 Gascon, F., Valle, M., Martos, R., Zafra, M., Morales, R., Castano, M. A., "Childhood obesity and hormonal abnormalities associated with cancer risk," *European Journal of Cancer Prevention*, vol. **13**, no. 3, pp. 193–197, June 2004.

20 Li, J., Fu, A. W., Fahey, P., "Efficient discovery of risk patterns in medical data," *Artificial Intelligence in Medicine*, vol. **45**, no. 1, pp. 77–89, January 2009.

21 Tsipouras, M. G., Exarchos, T. P., Fotiadis, D. I., Kotsia, A. P., Vakalis, K. V., Naka, K. K., Michalis, L. K., "Automated diagnosis of coronary artery disease based on data mining and fuzzy modeling," *IEEE Transactions on Biomedical Engineering*, vol. **12**, no. 4, pp. 447–458, July 2008.

22 Huang, Y., McCullagh, P., Black, N., Harper, R., "Feature selection and classification model construction on type 2 diabetic patients' data," *Artificial Intelligence in Medicine*, vol. **41**, no. 3, pp. 251–262, November 2007.

23 Pang, B., Lee, L., "Opinion mining and sentiment analysis," *Foundations and Trends in Information Retrieval*, vol. **2**, no. 1–2, pp. 1–135, 2008.

24 Ravia, K., Ravi, V., "A survey on opinion mining and sentiment analysis: tasks, approaches and applications," *Knowledge-Based Systems*, vol. **89**, pp. 14–46, November 2015.

25 Pang, B., Lee, L., Vaithyanathan, S., "Thumbs up?: sentiment classification using machine learning techniques," Proceedings of the Conference on Empirical Methods in Natural Language Processing (EMNLP-2002), Philadelphia, PA, July 6–7, 2002.

26 Moraes, R., Valiati, J. F., Gaviao Neto, W. P., "Document-level sentiment classification: an empirical comparison between SVM and ANN," *Expert Systems with Applications*, vol. **40**, no. 2, 1, pp. 621–633, February 2013.

27 Ghiassi, M., Skinner, J., Zimbra, D., "Twitter brand sentiment analysis: a hybrid system using n-gram analysis and dynamic artificial neural network," *Expert Systems with Applications*, vol. **40**, no. 16, pp. 6266–6282, November 2013.

28 König, I., Auerbach, J., Gola, D., Held, E., Holzinger, E., Legault, M., Sun, R., Tintle, N., Yang, H., "Machine learning and data mining in complex genomic data—a review on the lessons learned in Genetic Analysis Workshop 19," *BMC Genetics*, vol. **17**, Suppl. 2, pp. 49–56, February 2016.

5

Human-Centric IoT Networks

Albena Mihovska[1], Ramjee Prasad[1,2,3], and Milica Pejanovic[4]

[1] Department of Electronic Systems, CTIF, Aalborg University, Aalborg, Denmark
[2] CTIF Global Capsule (CGC), Rome, Italy
[3] School of Business and Social Sciences, Aarhus University, Aarhus, Denmark
[4] Faculty of Electrical Engineering, University of Montenegro, Podgorica, Montenegro

5.1   Introduction

The current Internet of things (IoT) concept is characterized with billions and billions of devices interworking through a myriad of technologies for the delivery of smart personalized services and applications. At the center of these is the human user who drives his/her own interconnected cluster. It can be expected in the future that the number of such clusters will grow exponentially, leading to an ultradense environment of interconnected devices belonging to the same or different clusters with the human user as the center point for the information being sensed, gathered, and processed. This concept has also been introduced elsewhere as "human center sensing (HCS)." The vision pushed by the authors of this chapter for the IoT beyond 2050 is one of the large-scale and dense HCS connectivities, the complexity of which can be used to extract new IoT value proposition.

IoT is a prominent information and communication technologies (ICT) paradigm, which enables information flows from/to and among highly distributed, heterogeneous, real, and virtual devices (sensors, actuators, smart devices). Although the IoT concept originated to enable communication among various types of devices (i.e., physical objects) for the provision of smart applications, IoT has evolved to enable forms of collaboration and communication between people and things, that is, to further evolve as an enabler of human-bond communications. From a technical point of view, the IoT is defined as the Internet of connecting the human and things with identifiers and/or information processing capabilities. Comparing with the existing networks, the IoT has the following significant characteristics: connecting directly with the physical world without human intervention, autonomic networking of the IoT nodes, and autonomic interaction between the IoT nodes.

In an IoT scenario, ICTs are merged with traditional infrastructures to serve as platforms for the gathering of data that can be utilized to deliver personalized services to the user. Although a subset of an IoT entails machine-to-machine (M2M) communications and applications, IoT applications have evolved to be highly personalized and to have the human user, and not the devices, as the driver of the IoT scenario dynamics. A typical IoT scenario entails a successful and cooperative interaction among things and individual users in an environment known as the "smart home" (also referred to as Ambient Assisted Living (AAL)) and expands beyond to the environment of a smart building and smart city. Such cooperative interaction is enabled by a large number of heterogeneous geographically distributed sensors and Internet-enabled devices, collecting and transmitting data to be real-time processed for the delivering of smart and personalized IoT applications. As a result, an IoT scenario needs to handle many hundreds (sometimes thousands) of sensor streams. At the center of the smart-home IoT scenario is the human who is the determining factor for the number, purpose, direction, and frequency of the sensor streams. Therefore, we refer to this scenario as an HCS scenario.

Jointly, with the benefits of such a technology arise a number of critical challenges, namely, the fact that the HCS elements and infrastructures are highly heterogeneous (e.g., sensors, RFID, smartphones, etc.), location specific, and resource constrained [1, 2]. In addition, with the continuous growth of Internet-enabled devices [3] comes the challenge of complexity and scalability of the HCS-based IoT networks that would grow exponentially with the growth of Internet-enabled devices. These challenges require a new approach to modeling the IoT networks that also reflect on the dynamics of the topology that comes from the unpredictability of the possible IoT connectivity links that may occur in a period of time. This obviously makes the network topology highly variable, and therefore, the currently proposed models are not able to reflect correctly on the resulting implications in terms of the efficiency of the resource provisioning, routing, quality of service (QoS), and security mechanisms, to mention the most critical ones.

This chapter proposes a new approach to dealing with the expected complexity of beyond 2050 IoT networks based on the concept of macro-simplicity. We propose a probabilistic macrolevel model that abstracts the complexity of the underlying microelements, such as the heterogeneity of the interconnected devices and wireless technologies employed, and the dynamics of the network topology and scenarios to derive a high level model that builds on simple macrolevel laws but sufficiently addresses all the micro characteristics of the complex IoT networks.

The chapter is organized as follows. Section 5.2 provides a research overview analysis of currently proposed models for IoT and summarizes their advantages and disadvantages on the backdrop of the HCS IoT scenario. Possible HCS scenarios demanding a change in the modeling approach will be described. Section 5.3 proposes the macrolevel model and describes how it can help to achieve simplicity in the complex IoT world. Our model is based on the enion probability analysis proposed by the author in [4]. This is a method for understanding how to get from micro-complexity to macro-simplicity and that allows for describing the behavior of the basic units of a complex system probabilistically. In the context of HCS IoT, the basic unit driving the behavior of an HCS cluster, which can comprise of various devices capable of transmitting and/or receiving information over the network, is the human. There are various microvariables, associated with an HCS cluster (e.g., its location). The goal is to allow an abstraction of microvariables by describing the system away from the individual things at a macrolevel by means of the enion probability analysis. We then describe the steps of this analysis performed for an IoT system. Section 5.4 concludes the chapter.

5.2 Overview of State of the Art in HCS Internet of Things

5.2.1 Human-Centric Sensing Networks and Federations (HCS-Ns and HCS-NFs)

The EU projects MAGNET and MAGNET Beyond, which were funded under the Framework Program 6 ICT, were the pioneer projects that brought about the concept of the personal network (PN) and personal network federation (PN-F) [5]. The PN concept went a step further than the initially known concept of the personal area network (PAN) [6], to define the virtual environment of a user spanning a number of connected devices and providing various personalized services. The same projects expanded the PN to the PN-F concept, which reflected on a scenario when the PNs of several individual users would interconnect in a cooperative way for the sharing of information and services. This research goes a step beyond the state of the art of the PN and PN-F to propose the concept of the HCS networks (HCS-Ns) and HCS-N federations (HCS-NFs).

5.2.1.1 HCS-N

An HCS-N supports the daily life of citizens in an unobtrusive way based on their personalized needs and thus is an enabling element of the AAL scenario.

The smallest unit of the HCS-N is the smart body area network (S-BAN). An S-BAN, as shown in Figure 5.1, will involve some wearable technology

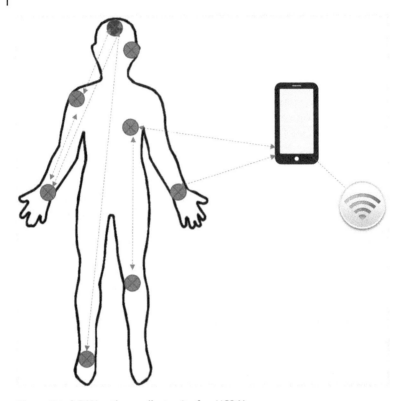

Figure 5.1 S-BAN as the smallest unit of an HCS-N.

(e.g., Fitbit, smart watch, etc.) and implantable technology (in support of restoring the sensory functions after spinal cord injury or similar). The wireless communication in an S-BAN requires low transmission power and miniaturized form factor that can support high data rates [8].

The dynamics and complexity of the S-BAN topology is determined by the user needs. A healthy individual may use only one or two wearable devices (sensors, smart watch) to monitor his/her sports performance (e.g., the Adidas miCoach sensor shoe technology) or out of curiosity (e.g., Fitbit). An elderly or a chronically ill patient (e.g., suffering from asthma, COPD, diabetes, and heart disease or in rehabilitation after a brain hemorrhage or a spinal cord injury) would require a more elaborate monitoring, thereby increasing the complexity of the personal S-BAN. In an AAL scenario, there would be a minimum of one S-BAN, and the maximum number of possible S-BANs will depend on the number of individuals forming a household, while the expected individual S-BAN complexity will be dependent on the health status of each AAL household member. It would be safe to assume that in a household of two healthy adults

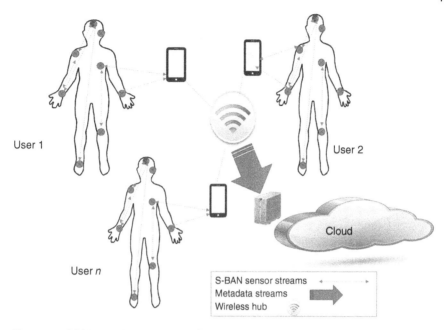

Figure 5.2 S-BANs coexisting in a smart-home scenario.

and their healthy children, the S-BAN will be of minimum and even in some cases zero complexity (if the users are not interested in daily statistics about their physical activity). In a smart-home or in-building (e.g., hospital) environment, it can be expected that several S-BANs will be coexisting and cooperating. In fact, such scenario should also be considered for an outdoor environment [9]. An S-BAN coexistence scenario is shown in Figure 5.2.

The smartphone collects the meaningful data aggregations from the body sensors. Data interaction may be initiated directly by the user (e.g., to check the real-time step counts and the percentage of the physical activity goal completion) or by the AAL system (e.g., to trigger alarms based on real-time processing of the aggregated data). The second type of interaction would require a real-time communication link to external (i.e., secondary) users enabling two sets of functionalities, namely, service and parameter configuration and complex monitoring [11].

A real-time signal processing node, also known as a "fog node" (in Figure 5.2, the smartphone can be the fog node), would receive and process the data collected by the sensors and delivered as sensor streams. Such data would be used for immediate reasoning and decision taking about the user's health status. Some data would be collected periodically to be delivered via a wireless hub and a gateway, as metadata stream to the cloud where it can be used for building personalized user applications and retrieved on demand [11]. The scenario of Figure 5.2 allows for the S-BANs also to cooperate for the purpose of resource and common information sharing. Cooperating S-BANs will be formed on demand, as the opportunity and need arises to support a user's personal applications. Therefore, an S-BAN may comprise of personal and foreign devices, interconnected through various communication technologies, which is what we term as the HCS-N.

The human user is at the center of the HCS-N, and thus, the complexity and topology of the HCS-N will also be determined by the individual user needs and mobility patterns. Because the user will always have his basic S-BAN as the smallest HCS network around him, we may freely assume that the S-BAN is the basic HCS-N unit, and, thus, the minimum complexity of the HCS-N of a user will be determined by the complexity of the S-BAN of that user. The HCS-N is built around the user to provide him/her with access to personal services, applications, and content regardless of the location. In other words, when the S-BAN connects to a number of smart appliances (washers, dryers, refrigerators, etc.) or safety and security systems (Internet-connected sensors, monitors, cameras, and alarm systems) or energy equipment like smart thermostats and smart lighting, there will be an HCS-N formed. A nomadic user will build and own HCS-N that would always comprise the S-BAN and the devices that a user need to interconnect to depending on the personal needs in a particular location. This includes a scenario of interconnecting to certain devices remotely over the communication infrastructure similar to the way a PN would connect [5]. Thus, an HCS-N will be of dynamic nature, in terms of connectivity, composition, and configuration.

There are a number of enabling technologies needed to realize the HCS-N scenario. These are the following:

Energy-Efficient Communication Technologies for Interference-Free Intra- and Inter-S-BAN Connectivity An S-BAN will require that a number of heterogeneous devices operating on various radio frequencies and with different operational characteristics coexist and communicate. Typically, connectivity among the elements of an S-BAN is enabled by short-range wireless technologies, such as WLAN, Zigbee, Bluetooth low energy (BLE), and ultra-wideband (UWB). The European Telecommunication Standardization Institute (ETSI) [9] carried out coexistence measurements and analysis in an effort to specify the requirements for the S-BAN devices for three types of indoor coexistence scenarios, namely, hospital, home, and office.

One challenge related to S-BAN connectivity is driven by the fact that most of the devices would be operating in an unlicensed band. The advent of cognitive radio supposes that these devices would be able on occasions to also use unoccupied licensed spectrum. This has intensified the research toward use of non-RF technologies, such as visible light communications (VLC) [7, 12], making it another S-BAN connectivity enabler. VLC has the potential to release the burden of over-occupied licensed and unlicensed RF spectrum and has the additional advantage of being security resilient, which is extremely important for an S-BAN scenario. Another open issue is the need for radio channel modeling that reflects the effect of the human body and tissues on the RF signal propagation in order to enable efficient and reliable transmission of vital data [8, 13]. Research reported in Ref. [8] showed that the impulse response of a signal traveling through the human body differentiates from the specifications of the IEEE 802.15.4a WPAN standard [14], which would mean that some S-BAN scenarios (e.g., implantable devices) will not be feasible when building on this technology. Therefore, the promising open research issues for enabling reliable S-BAN connectivity are toward RF interference mitigating technologies, energy-efficient short-range technologies, new channel models to account for the specifics of the human body medium and their effect on the S-BAN radio transmissions [8, 13], and a seamless handover between RF and VLC transmission technologies. Some initial proposals for integrating VLC into the AAL scenario were presented in Ref. [7].

Indoor Localization and Positioning Indoor localization and positioning are an essential part of the HCS-N scenario and a key enabler of the HCS-N services and applications. As an example, the ability to track the movements of a remotely cared-for patient, located at home, and especially in critical situations (e.g., falls), is paramount to enabling the full potential of eHealth systems and applications. However, this technology is still very much an open research field. The currently existing methods and algorithms for indoor localization are based on some sort of wearable devices or environmental sensors that detect the peoples' position. The most commonly used techniques are trilateration, triangulation, time of arrival, and fingerprint [15]. Mandel and Autexier [10] investigated the use of low-cost thermal imaging sensors for the application of people tracking in a smart-home environment. Palumbo and Barsocchi [16] proposed to fuse the information (coordinates) provided by a localization system with the information coming from the binary sensor network deployed within an AAL environment. BLE has recently gained a lot of attention because of its low power consumption, reasonable price, and ease of use, and in addition it is already an implemented feature of every smartphone. Mitev and Stoyanova [17] investigated the achievable accuracy for BLE enhanced with indoor positioning of fallen patients for several usage scenarios, where the received signal strength indicator (RSSI) was a decisive metric. It was shown that the usage scenario specifics would impact the optimal choice of an indoor positioning technique (e.g., fingerprinting, triangulation) and, thus, would have a strong effect on the achievable accuracy. Indoor localization is key for enabling cooperation and coexistence of S-BANs because several S-BANs in the same AAL environment may be interconnecting with the smart appliances and devices in order to provide personalized to the user services and applications. Here, it should be added that context awareness will be key to enabling the aforementioned.

Security, Privacy, and Trust, Including Identity Management (IdM)   The focal point of the HCS-N is the human user. The analysis of the user's needs and behaviors is essential for a human-centric operation; however, such an analysis is built on a lot of personal data that require to be stored somewhere. The acceptability of the HCS-N system depends on the risks felt by the users, the public authorities, and the policymakers. In certain S-BAN scenarios (e.g., eHealth use cases), attackers may use the personal data for harmful purposes.

Security threats and attacks in an HCS-N scenario can be classified as passive and active. For example, a passive attack may occur after the sensitive personal data gets routed through the network. Attackers may change their destination or make routing inconsistent. User-centric data may get stolen by eavesdropping to the wireless communication media. An active attack may occur if the location of the user becomes known, which may lead to life-threatening situations. The common design of sensor devices incorporates limited external security features and, therefore, makes them prone to physical tempering. This increases the vulnerability of the devices and poses more complex security challenges. The attacks to an HCS-N would most likely be due to eavesdropping and modification of the sensor data, allowing to trigger false alarms, denial of service, location and activity tracking of users, physical tampering with devices, and jamming attacks [18].

Such risks have legal, ethical, and technical dimensions. One possible countermeasure is the proper user identification and putting in place intelligent IdM mechanisms, which involve all of the HCS-N elements [19]. The identification can be explained as an association of attributes, which represents identifiers. To make the communication within the HCS-N more efficient, one common device and object identifier would be beneficial [20–22]. Zdravkova *et al.* [19] proposed a computing device recognition (CDR) algorithm for automated and secure user and device identification to be applied in an HCS-N eHealth scenario. It allowed that the user could be identified and access various service delivered by heterogeneous devices. Another advantage is the ability for a user to define and manage his/her own rules in the stage of profile creation such as device connectivity type (automatically or manually), services access permissions, and number of computing devices as an input data for CDR procedure.

Privacy-protecting technical concepts are key to an HCS-N, where diverse user-centric decisions are based on the inputs received from the various sensing and monitoring devices comprising the HCS-N. In addition, ethical requirements are also be very relevant here, particularly for HCS-N scenarios involving secondary users (e.g., medical personnel, family caretakers).

According to the EU and international law, people have the right to protection of their privacy and personal data. Personal data must be processed fairly for specified purposes and on the basis of the consent of the person concerned or some other legitimate basis laid down by law. Data protection law is made up of a broadly recognized set of principles: lawfulness, profiling prohibition, data availability, purpose limitation, data security, data subject rights, anonymity, responsibility, and accountability. The right to privacy includes the right to control personal data. This means that the user must be aware of the data and the time period for which their personal data are stored and who gets access to the information. Further, the user has the right to object to the data processing. Keeping this in mind, the funded under the European Framework Program 7 (FP7) Project eWALL proposed the privacy-by-design (PbD) approach [18]. The PbD was proposed for an eHealth system to enable the aspects of products and services, whether standardized or specialized, for any personal identification or behavioral profiling purpose, offered by equipment providers, equipment integrators, system integrators, and service providers. The approach should relieve caregivers and other users of any additional bureaucratic legal procedures that the technology deployment would imply. The PbD approach is very relevant for an HCS-N scenario and should be explored further.

Fog and Cloud Computing for Real-Time and Non-real-time Data Processing and Storage In an HCS-N scenario, data would be generated and collected from sensors and would be delivered to a dedicated storage area. The issue arises when these data start to grow, especially in the context of an interconnected HCS-N scenario. Fog computing allows for processing and storing real-time sensor data required to trigger alarms and similar applications requiring fast response, while the rest of the metadata streams can be processed in the cloud as shown in Figure 5.2. Another important capability to be taken in account here is the simultaneous processing of data from multiple sources. To this end one fog node should be able to use the resources of other fog nodes in the vicinity to cope with the possible explosive growth of the collected data.

Software-Defined Networking (SDN) An HCS-N is highly heterogeneous, and in addition, the human-centricity demands properties, such as modularity and adaptability. SDN is crucial to abstracting the heterogeneity of the devices comprising an S-BAN, allowing for customization and service flexibility. SDN allows for achieving differentiated quality levels to different HCS-N tasks, thus, enabling a common platform for integrating the HCS-N heterogeneous communication and networking technologies [21]. Modularity is essential to enable flexibility in terms of devices and the data these would collect that can enable the HSC-N services and to easily integrate novel devices into an existing HCS-N system. Adaptability would in turn provide for an ability to scale the HCS-N to the specific use cases.

Haptic Communications Haptic communications has recently emerged due to the advancements in the area of virtual and augmented reality and can be seen as an enabler of human-bond communications due to the presence of a strong perceptual element [23]. It is a field that is still relatively unexplored but can be related in essence to the impact of audiovisual presence, allowing for a user to interact with the environment through all the senses. Haptic communications build upon ultrafast response times, ultralow latency (<1 ms end-to-end), and carrier-grade reliability to enable haptic and tactile sensations [3].

5.2.1.2 HCS-N Federations (HCS-NFs)

In an HCS-NF, the HCS-Ns of different users will cooperate for a particular purpose and would be able to share resources and information. The concept is similar to the cooperating S-BANs in Figure 5.2; however, a HCS-NF would have a larger span, across cities, countries, and continents, and would be formed on demand, thus displaying a highly variable and context-driven topology. This cooperation will be supported by the larger ICT infrastructure, and would be typical for a smart city scenario. The requirements defined for the HCS-N will generally stay the same, increasing the complexity of the required solutions, especially the ones related to security and privacy because in the HCS-NF scenario, security and privacy must be maintained across geographically distributed HCS-Ns and their belonging personal devices, which can be seen as a multiuser virtual private network overlay [5].

The HCS-NF operation relies on the automatic management, identification, and use of a large number of heterogeneous physical and virtual objects (e.g., both physical and virtual representations), which are Internet-enabled and may be on various locations. A common aspect of HCS-NF applications is that they may involve control and monitoring functions, where human-in-the-loop actions are not required, which further tightens the security, privacy, and reliability requirements. A scenario of a HSC-NF is shown in Figure 5.3.

5.3 Modeling of HCS-Ns and HCS-NFs

The continuous growth of the world population would definitely influence the growth of the number of S-BANs and the variety of personal applications needed, while the continuous growth of interconnected devices will influence the complexity of an HCS-N and the HCS-NF, which is expected to increase rapidly [24]. In addition, a large part of the established connections will be autonomous and their formation will be based on local decisions determined by the user needs. If we think of the HCS-N connections in the scope of the complete IoT ecosystem, it can be assumed that each HCS-N will be to a degree dependent on the behavior of the HCS-Ns around it, leading to an intensely complex scenario.

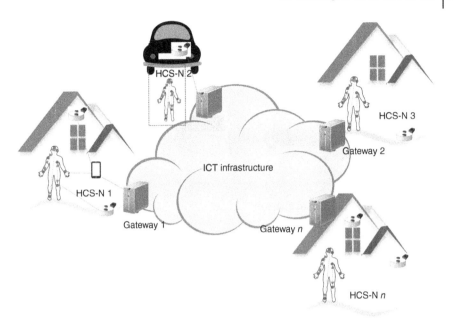

Figure 5.3 HCS-NF scenario.

Thus, we can expect that the existing models proposed for ad hoc connectivity will not be sustainable to the level of the expected HCS IoT system complexity.

Here, we seek to define and characterize the simple behavior of the HCS IoT system by means of a probabilistic macrolevel model that abstracts the complexity of the various possible heterogeneous connections and sub-scenarios within the context of HCS-N and HCS-NFs. The model is based on the *enion probability analysis*, proposed for modeling complex systems by Michael Strevens [4]. The goal is to define the macrolevel law, describing the simple dynamics of the cooperation of the HCS-Ns and HCS-NFs within an IoT ecosystem and that could be expressed by means of only a few variables.

In the context of HCS IoT, the basic unit or the "enion" is the S-BAN. An S-BAN may comprise of various heterogeneous devices capable of transmitting and /or receiving information over the network. The way the various S-BANs change topology or interact with one another would drive the behavior of an HCS-N and thus its complexity. The way an S-BAN will interact within an HCS-N will be based on a human-centric decision and dependent on the personal application needs. Therefore, the topology of a possible HCS-N will change constantly, depending on the human-centric needs. In a cooperative HCS-N scenario, it would be the human-centric needs that would determine whether and how individual HCS-Ns will interconnect, which could be in an

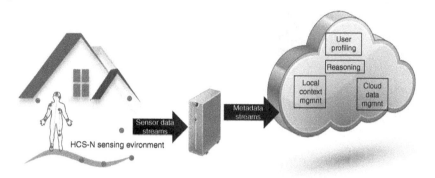

Figure 5.4 HCS-N supporting platform.

ad hoc manner or over the existing infrastructure, and the possible connections over which information would be routed will be highly variable. Still, there will always be a basic set of personalized applications associated with a single HCS-N. Thus, the probabilistic element defining the HCS-N dynamics will be the *user activity profile* associated with S-BAN.

Figure 5.4 shows an HCS-N scenario with the supporting platform for the delivery of the human-centric applications. In a typical HCS-N, the information gathered would be semantically enriched, aggregated, and fused to provide higher level context data that can be used for activity and situation recognition. Such data can be gathered from the immediate environment (e.g., sensors, A/V devices, etc.) or from external sources (e.g., social networks, web feeds, etc.). The sensing environment collects various data about the user and the environment from the medical and environmental sensors and actuators. The data from the human-centric environment will be configured, queried, and reported to the data management blocks located in the cloud for further storage and processing.

The local context management would handle low-level context data and processing (e.g., indexing and adaptation, analysis of humans, and non-A/V perception processing).

The user activity profile will be based on information about the user's lifestyle and health condition/needs. For this purpose, a lifestyle reasoner functionality has been proposed to be integrated to process and store long-term data related to certain patterns or routines that define the lifestyle of the user [11]. The lifestyle reasoners consume medium-level data (e.g., data used to generate the services) from multiple sources and derive semantically meaningful patterns. The data is processed, stored, and compared with medium- and long-term data stored in the cloud, and the reasoner would determine whether a variation falls within the expected thresholds or would employ more complex methods to determine some deviation from the usual behavior that may trigger a possible alarm. These results may be exposed through an API for use in other applications or other processing components. The reasoner in Figure 5.4 would make decisions about the short-, medium-, or long-term user events and behavior. The implementation of various lifestyle reasoners would be the highest level of data processing within an HCS-N supporting platform.

The topology and the expected interactions of an HCS-N, therefore, will depend on the daily activities of the user, which in turn will be built around an *a priori* known as user's lifestyle and health state pattern. Therefore, the law governing the complexity of the HCS IoT would owe its simplicity to the probability that a given HCS-N, which is a part of the HCS IoT ecosystem, would follow the predicted stable state of a particular user activity profile. The simple behavior of the HCS IoT system will be the cumulative consequence of the probabilistic behavior of the HCS-Ns constituting the IoT system. This probabilistic behavior can be expressed with only a few variables, for example, is an HCS-N behaving according to the specifics of the user activity profile or is it deviating from those? Or how many alarms related to a given user profile have been triggered?

A critical issue related to the design of an HCS IoT system model based on the user activity profiles would be to effectively enable privacy and data protection. To build a model based on a user activity would require a protection mechanism that would not allow for linking the behavioral or location data of the users with databases for secondary purposes nor for extracting sensitive health data about a user without authorization. A pseudoidentifier (PI) allows for depersonalizing the data, which minimizes the risks of data collection and data sharing. Within the context of eHealth systems, it has been proposed to add a mechanism to split up the data that can be used to identify patients into PIs and auxiliary data (AD) [25]. In addition, users should be able to choose themselves the degree to which the HCS-N system components would involve the use of PIs. The larger the number of PIs, the higher the level of the privacy protection. Further, the model should allow for deleting the behavioral raw data or switch off certain sensors (cameras, GPS, accelerometer, touch screen, oxygen sensor, bed and door sensors, etc.) and separate metadata from actual features while maintaining full functionality for the purpose of preserving anonymity of the data. The PbD approach was proposed in Ref. [25] in relation to remote care for chronically ill patients and the elderly. There, a privacy-protecting habit extraction technical component was implemented to create PIs for the activities and situations used for sudden and slow deviation detection. If users choose for strict privacy, the system would transform the semantic models into representations to detect deviation from the combined lifestyle patterns. PbD will be very useful in the context of HCS IoT and the related services and applications.

Other technological requirements needed to be considered when building a model based on human-centricity, are related to context and profile, IdM and choosing an appropriate ontology framework that maybe common for all these.

The supporting platforms must enable extremely low latency in combination with high availability, reliability, and security to adequately respond to the dynamics of the HCS IoT system [3]. The concept of the tactile Internet is also a key enabler of HCS communications and networking and can benefit from a macrolevel approach to the HCS IoT model [26]. Another emerging field that supports the HCS IoT concept is the already mentioned area of haptic communications [23]. Haptic communications have the potential to enhance the information of audiovisual data collected by the devices within an HCS-N and provide for truly immersive HCS-N applications and services and for enriched multiuser interaction within HCS-NFs.

5.4 Conclusions

This chapter proposed a conceptual approach to IoT modeling that includes not only the sensors but also the future actuation and smart behavior that is key to enabling HCS applications and services. The increasing human-centricity of these applications alters the needs of the future IoT systems in terms of scalability, heterogeneity, complexity, and dynamicity. Therefore, an approach that enables full characterization of the system behavior but is based on only a few variables would be very valuable to meet those needs.

The expected HSC-N dynamics can be expressed through the associated user activity profile. Unexpected HCS-NF formations can be detected based on the deviations from the basic settings of a given user activity profile, but it is possible based on a lifestyle reasoning approach to probabilistically predict the dynamics of the HCS-NF around the personal HCS-N.

More research is needed to introduce the required technologies for enabling end-to-end security and privacy and data protection that are key to the proposed approach. Another important open area of research is related to identity and context management. The emerging fields of haptic communications and the tactile Internet are very strong enablers of the HCS-N and HCS-NF interactions and have the potential to open the field toward numerous innovative HCS IoT usage scenarios.

Finally, HCS has the potential to bring innovation and added value to the current ICT infrastructure. It would enable to introduce the individual user into the services and applications business model by opening up opportunities for creating new ones based on individually generated content. Modeling the HCS system around the user activity profile provides a good basis for enabling the aforementioned.

References

1 K. Lee and D. Hughes, "System architecture directions for tangible cloud computing," In *International Workshop on Information Security and Applications (IWISA 2010)*, Qinhuangdao, China, October 22–25, 2010.

2 K. Lee, "Extending sensor networks into the cloud using Amazon web services," In *IEEE International Conference on Networked Embedded Systems for Enterprise Applications*, November 2010, Suzhou, China.

3 ITU-T, "The tactile Internet," Technology Watch Report. Available at: http://www.itu.int/en/ITU-T/techwatch/Pages/tactile-internet.aspx (accessed October 2016), August 2014.

4 M. Strevens, *Bigger than Chaos: Understanding Complexity through Probability*, Cambridge: Harvard University Press, September 2006.

5 R. Prasad (Ed.), *My Personal Adaptive Global Net (MAGNET)*, ISSN 1860-4862, Dordrecht, Heidelberg, London, New York: Springer, 2010.

6 R. Prasad (Ed.), *Towards the Wireless Information Society*, Volume II, Norwood: Artech House, 2006.

7 A. Kumar, A. Mihovska, S. Kyriazakos and R. Prasad, "Visible light communications (VLC) for ambient assisted living," *Springer International Journal on Wireless Personal Communications*, vol. 78, no. 3, pp. 1699–1717, October 2014.

8 P.V. Patel et al., "Channel modelling based on statistical analysis for brain–computer-interface (BCI) applications," In *Proceedings of IEEE INFOCOM*, San Francisco, CA, April 2016.

9 ETSI Smart-BAN, Draft V0.1.0 (2015-10), "Measurements and modelling of SmartBAN RF environment," Technical Report, 2015, ETSI online. https://portal.etsi.org/TBSiteMap/SmartBAN/SmartbanToR.aspx

10 C. Mandel and S. Autexier, "People tracking in ambient assisted living environments using low-cost thermal image cameras," In C.K. Chang et al. (Eds.): *Inclusive Smart Cities and Digital Health*, ICOST 2016, LNCS 9677, pp. 14–26, Basel: Springer International Publishing Switzerland, 2016.

11 EU FP7 ICT Project eWALL, D2.7, "Final user and system requirements and architecture." Available at: http://ewallproject.eu (accessed October 20, 2016), March 2015.

12 R. Prasad et al., "Comparative overview of UWB and VLC for data-intensive and security-sensitive applications," In *Proceedings of IEEE ICUW 2012*, Syracuse, NY, pp. 41–45, ISSN 2162-6588, September 2012.

13 R. Prasad and A. Mihovska, *New Horizons in Mobile and Wireless Communications, Volume 1: Radio Interfaces*, Norwood: Artech House, 2009.

14 IEEE 802.15 Webmaster, "IEEE 802.15.4a WPAN Standard." Available at: http://www.ieee802.org/15/pub/TG4a.html (accessed October 20, 2016), 2016.

15 M. Al-Ammar et al., "Comparative survey of indoor positioning technologies, techniques, and algorithms," In *International Conference on Cyberworlds*, October 2014, Santander, Spain.

16 F. Palumbo and P. Barsocchi, "SALT: source-agnostic localization technique based on context data from binary sensor networks," In E. Aarts et al. (Eds.): *Ambient Intelligence*, LNCS 8850, pp. 17–32, Basel: Springer International Publishing Switzerland, 2014.

17 M. Mitev and S. Stoyanova, "Indoor positioning for smart ambient assisted living services." M.Sc. Thesis, Department of Electronic Systems, Aalborg University, 2016.

18 EU FP7 ICT Project eWALL, D2.4, "Ethics, privacy and security." Available at: http://ewallproject.eu (accessed October 20, 2016), April 2014.

19 V. Zdravkova et al., "Single thing sign on identity management for Internet of Things," In *Proceedings of Wireless VITAE 2015*, Hyderabad, India, December 2015.

20 D. Todorov, *Mechanics of User Identification and Authentication: Fundamentals of Identity Management*, Boca Raton, FL: Auerbach Publications/CRC Press, 2007.

21 Z. Qin et al., "A software defined networking architecture for the Internet-of-Things," In *Proceedings of IEEE NOMS*, Krakow, Poland, May 2014.

22 P. Mahalle, N. R. Prasad and R. Prasad, "Identity management framework towards Internet of Things." PhD Thesis, Aalborg University, November 2013.

23 E. Steinbach et al., "Haptic communications," *Proceedings of the IEEE*, vol. **100**, no. 4, pp. 937–956, April 2012, doi: 10.1109/JPROC.2011.2182100.

24 S. Laghari and M.A. Niazi, "Modeling the Internet of Things, self organizing and other complex adaptive communication networks: a cognitive agent-based computing approach," *PLoS One*, vol. **11**, p. e0146760, 2016.

25 EU FP7 ICT Project eWALL, D2.8, "Report on the privacy-by-design approach." Available at: http://ewallproject.eu (accessed October 20, 2016), July 2014.

26 A. Aijaz et al., "Realizing the tactile Internet: haptic communications over next generation 5G cellular networks," In *Proceedings of IEEE Wireless Communications and Networking Conference Workshops (WCNCW)*, Oulu, Finland, April 2016, doi:10.1109/WCNCW.2016.7552676.

6

Body as a Network Node: Key is the Oral Cavity

Marina Ruggieri and Gianpaolo Sannino

Center for Teleinfrastructures (I-CTIF), University of Rome "Tor Vergata", Rome, Italy

6.1 Introduction

Human bond communications (HBC) challenges the tight, synergic, and effective cooperation between the medical (bio, clinic, and surgery) area and the ICT engineering one. The interdisciplinary Italian CTIF (I-CTIF) has been established with the vision of putting the two areas in an osmotic interaction in order to be ready for the HBC challenge (e.g., [1]).

The human body will be the main actor in the ICT systems, by playing an *active* role as node of the ICT network as well as part of the ICT user terminal.

Body as a node (**ByN**) is the future of ICT networks.

On the other hand, "intrusion" with technological ICT devices in the body will provide to the body itself a great opportunity from a medical viewpoint (e.g., [2, 3]).

To this respect, in the last 15 years, the introduction and increasing application of wireless body area networks (WBAN) has provided to the user the familiarity with a set of portable (e.g., in a bag, in the pocket, or by hand) and wearable (e.g., shoes, glasses, clothes) devices that may be even implanted in or surface-mounted on the body with proper miniaturization and biocompatibility performance [4 and references therein, 5].

The WBAN-based system is mainly a sensor network composed of body sensors, with various miniaturization levels, and a central body-related hub.

In those networks larger-size devices (like smartphones and tablets) still play a role in offering a user-friendly interface for data management and a bridge from the short-range network around the body to long-range networks.

Human Bond Communication: The Holy Grail of Holistic Communication and Immersive Experience, First Edition. Edited by Sudhir Dixit and Ramjee Prasad.

The core applications of the previous approach lie mainly in the healthcare realm, both in codified medical doctor–patient exchanges or in amateur and sport health-related sensing.

Microsensors and small sensors are the key hardware elements in the body-related sections of the WBAN-based networks.

The previous frame highlights a core concept underlying the WBAN scenario: the body is related (via gateway) to external networks where, in particular, data transfer is implemented.

The role of the body, except for the capability and willingness of carrying, wearing, and, eventually, embedding (implant or surface-mounted) sensors, is *passive* from the network function point of view.

The ByN concept is different and it could change the way of conceiving, designing, and implementing ICT networks in the future.

In the ByN network, the body becomes *active* and *fully integrated* in the network itself.

In a long-term vision, the body will not belong to a user of the network, but it will be part of the network, concurring in both the local data handling, through a bio-personalized user interface, to the data transfer, and to additional functions that might be needed.

A non-harmful bio-optimal engineered solution is the final landing point [6].

In this framework, the entry gate of the ByN approach is, in authors' opinion, the oral cavity (OC). In particular, OC is envisaged as the main and most effective entry point in the emulation and experimental phases, as well as one of the key interface elements in the final implementation phase.

In order to create the oral cavity as a node (OCN), the medical experts of the team have to be both great professionals and brave researchers [7, 8]. On the other hand, the ICT experts have to believe deeply that the interdisciplinary approach is the key for the benefit of technology to mankind [9].

The chapter will describe the ByN approach and specialize it to the OC, highlighting the future perspectives and implications of the envisaged outcomes.

6.2 The Body as a Node Approach

The role played by the user in future mobile networks is changing. At present, the transformation rate seems "reasonable" and tightly related to the technological progress of network and devices and to the social changes inside the human society.

Nonetheless, the transformation that the mobile technology has operated to the society is much deeper and broad we can appreciate so far. The digital approach is pervading everyday life of an enormous number of people in a "democratic" way that preserves gender, age, and geographic diversity.

With a strong statement we could affirm that human beings are getting more and more "digital inside." Obviously the sentence refers to a digitalization that is a "state of mind" and a potential availability to a technological progress that may transform the theoretical inner digitalization into something of a more practical meaning.

In what follows the latter framework will be referred as "***state D***" (***D*** as inner **D**igitalization).

Furthermore, many people believe that the tremendous technological progress of the digital world is not translating at the same rate to a medical progress that is able to prevent bodies from suffering from dangerous and, in many times, mortal diseases.

In what follows the latter framework will be referred as "***state M***" (***M*** as **M**edical relief).

The proper harmonization between states D and M is the aim of the research authors' are undertaking.

In order to translate state D from a theoretical state to a partial implementation, the starting point is an increasing synergy between the mobile user and his/her device and, through it, with the network. The next frontier is a body-related device that will provide friendly solutions to the intrinsic limitations of the current interaction of the human body (e.g., fingers) and the device (e.g., keyboard) [10].

Sense augmentation through the devices is already in progress [10–13]. However, senses are only one aspect of the human body. The human body is much more than that.

A further step could be a strong interaction between the user's brain functions and the device. A possible scenario would envisage the device be driven by brain and emotion-related commands. In the depicted frame, the device would need not only for pure technology but also for a proper merging between standardization and ethics (*stethics*) [10].

The previous steps are a possible way to decrease the conceptual and physical distance between the human body and the user device (***BD distance***), but they are not the final answer and they are not mandatory to reach the aimed joint synergy between states D and M, which would allow to create a new final state (***state DM***).

The transformation of state D to practical implementation envisages a progressive "intrusion" in the body with technological ICT devices.

The approach could gradually bring the human body of the user to become cooperative with the network and the user device, internalizing part of the functions of a network node (*ByN*).

If $\mathbf{F_{ND}}$ indicates the overall number functions that a network node currently implements, the ByN vision envisages to first identify those functions that in a medium-term vision can be fully softwarized ($\mathbf{F_S}$). A high degree of softwarization in network, node, and device functions is the key point to start properly the ByN implementation.

F_S is an increasing function of time, as the ongoing softwarization of network, node, and device functions is rapidly progressing [14, 15].

A thorough investigation from the joint group of ICT and medical experts will identify the group of functions F_{INT} that could be "internalized" in the human body. Those functions could be partly implemented through the body itself and partly by implanting external aids.

The remaining groups of functions F_{EH} are those needing an external hardware to be implemented.

Therefore, in summary, we can write

$$F_{ND} = F_S + F_{EH} + F_{INT} \qquad (6.1)$$

The conceptual block diagram is shown in Figure 6.1, where the **IAF** block indicates the analysis of functions performed through an interdisciplinary approach.

It is worth highlighting that the intrusion of external devices for medical purposes is already a well-accepted and implemented concept. Furthermore, biomedicine is progressing dramatically, and the early diagnostics of important diseases can take advantage of proper chips to be implanted in the body (e.g., [2, 3]).

To this respect the intrusion in the user body for the implementation of node functions F_{INT} can be effectively integrated by biochip functions F_{MED} related to early detection, monitoring, and progressive cure of special diseases.

In the previous frame, the overall internalized functions F_{body} would be given by

$$F_{body} = F_{INT} + F_{MED} \qquad (6.2)$$

A successful implementation of the ByN approach can gradually transform a "full technical" network (all technical **T** nodes) (Figure 6.2) into a network

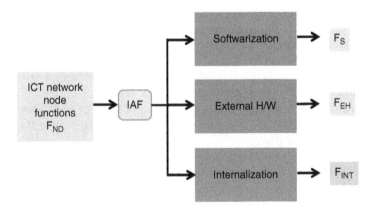

Figure 6.1 Function partitioning for ByN implementation.

Figure 6.2 Network with all technical T nodes.

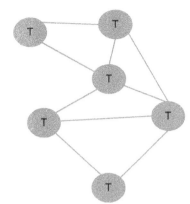

Figure 6.3 Network with a PH node in the implementation of the ByN approach.

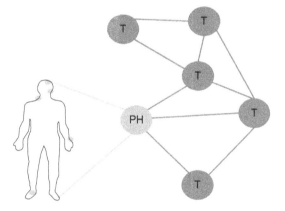

integrating T and ByN nodes, where F_{INT} of the node functions are implemented inside the body of the network user (**PH** nodes, **P**artial **H**uman body node) (Figure 6.3).

The next section will provide a possible answer to the question "where do we start for a ByN implementation?"

6.3 Oral Cavity as a Node

As highlighted in Section 6.2, the group of functions F_{INT} "internalized" in the human body could be partly implemented through the body itself and partly by implanting external aids. This hybrid solution (*type B*) represents an intermediate stage from a fully intrusive chip (*type A*) and a fully bio solution (*type C*).

In ByN, the body contribution could be distributed, as depicted in Figure 6.4. In the previous frame, the OC can play a key role. OC is particularly attractive in the emulation and experimental phase of the ByN implementation.

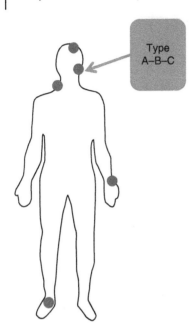

Figure 6.4 Body contribution to the ByN implementation, with focus on the OC.

Type
A–B–C

Figure 6.5 OCN approach.

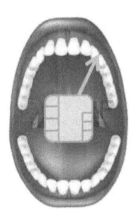

In fact, there is a natural availability to various degrees of intrusion inside the OC that ranges from dental fillings to orthodontic appliances up to implants. The OC is the place where people accept an "intrusion" for a long time or even for a permanent duration.

Furthermore, OC includes all the bioinformation about the body that can be necessary to implement the bio-optimal engineered ByN architecture.

In the previous frame, the ByN implementation can start from *OCN* approach (Figure 6.5).

On the other hand, the integration of the medical functions F_{MED} in the intrusive chip is particularly meaningful for the OC.

In fact, OC is one of the main gateways to the body. It is since the early months of life a body involved in carrying out different functions (immuno-protection, digestion, chewing, swallowing, phonetics).

The complexity and the sensitivity of these structures make oral health closely related to general health conditions. The first evidences of many systemic diseases are borne by the oral structures and therefore are diagnosed through an examination of those.

OC may, hence, constitute the ideal environment for positioning chips, including sensors, and collecting data that might be useful in monitoring oral activities, detecting pathologies/problems, and planning early intervention and partial remote management.

Due to the ICT contribution, performance and costs of cures for both the public health system and the private patients could change, especially in the dentistry field, where a significant growth of technology-driven approaches is occurring [7, 16]. However the potential opportunities offered by this sector have not been truly exploited yet, since these technological advancements have focused on the aspects of close dental relevance.

Using intraoral micro-biosensors featuring a wireless transmission could open new scenarios for dentistry applications (related to oral health), general sensing applications (sensing parameters related to human health), and human–machine interface (HMI) applications (using measurements of oral cavity to drive/control machines). Thanks to sensor miniaturization, the sensors can be easily integrated into prosthetic devices and orthodontic appliances, even in the teeth or bones, with minimum discomfort and without tissue injury [8].

Wireless communication between intraoral devices and the outside world is challenging, since the OC is a hybrid propagation environment. Intraoral sensors are in the border between intra-body and wearable sensors. It is not strictly speaking that a communication "inside the body" as a sensor inside the body, depending on its anatomical location, is usually surrounded by a distinct and fairly stable tissue environment.

An intraoral device (IOD) is located in a constantly changing environment depending on the relative positions of the jaws and movements of the tongue, which continuously changes shape when one swallows, breathes, or speaks [17–21].

As IODs are in contact with the gums, the tongue, or the palate, and their associated receivers may be in contact with the human body, and body channel communication (BCC) can be considered to be used for these devices [22].

However, when applied to IODs, neither of the BCC methods maintains their desired operating conditions due to attenuation effect. Therefore, the use of RF-based wireless communications could be considered. As a matter of fact, RF wireless communication has been already successfully adopted in implantable devices such as the pacemaker and neurostimulator. The proper selection of the frequency band is fundamental for establishing a reliable link between intraoral devices and the outside world. The selection of the frequency is strictly related to two fundamental design elements: size of the antenna and attenuation. It is worth noted that an RF-based communication approach foresees the use of a battery inside the mouth, at least for those important applications that foresee a continuous monitoring. This is one of the main challenges limiting the adoption of IODs for continuous monitoring. Interesting solutions for recharging the sensors are based on the use of the jaw movement that normally occurs when chewing, eating, and speaking [23].

For instance, one can obtain approximately 580 J only from daily chewing, which is equivalent to an average power of approximately 7 mW. An alternative approach has been proposed in two studies [24, 25] where a passive sensor is inserted in the mouth and read by an external device.

It should be highlighted that the presence of the sensors inside the OC might cause metabolic alterations (causing inflammation or tissue damage). Therefore, the molecular and biochemical approaches undertaken to unravel possible metabolic alterations (with particular reference to quantification of recognized markers of inflammation or tissue damage), occurring during prosthesis implantation and long-term wearing, should be addressed [26].

The presence of sensors and chips has to be, therefore, analyzed also in terms of potential harm to the oral cavity, in particular, and the whole body in general. The sensor/chip design and implementation have to be driven by the guidelines and requirements deriving from the previous analysis. The same applies to whatever part of the body is selected for the chip positioning.

6.4 Conclusions and Future Perspectives

The development of the *ByN* approach envisages a thorough analysis of the node functions that can be implemented in the human body through either type A or B or C solutions.

The ByN implementation will likely have OC as the key starting point, particularly in the emulation and experimental phase. The harmonization of both state D and M and the design in the medium term of an effective DM state are expected.

OC can play a key role also in the final implementation phase, where the body interface points have to be selected.

The chapter has presented the theoretical and systemic setting of the ByN approach and the possible issues and opportunities. A long-term implementation envisages a non-harmful (for the human body) bio-optimal engineered solution.

A solid interdisciplinary approach and balance among the various elements and challenges are crucial aspects for the effective future implementation stage.

Acknowledgments

Authors wish to acknowledge Prof. Massimo Coletta and Prof. Ramjee Prasad, both from CTIF, for the useful discussions.

References

1 Sannino, G., Sbardella, D., Cianca, E., Ruggieri, M., Coletta, M., Prasad, R., "Dental and biological aspects for the design of an integrated wireless warning system for implant supported prostheses: a possible approach," *Wireless Personal Communications* 2016; **88**(1): 85–96.

2 Warkiani, M.E., Khoo, B.L., Wu, L., Tay, A.K., Bhagat, A.A., Han, J., Lim, C.T., "Ultra-fast, label-free isolation of circulating tumor cells from blood using spiral microfluidics," *Nature Protocols* 2016; **11**(1): 134–148.

3 Warkiani, M.E., Guan, G., Luan, K.B., Lee, W.C., Bhagat, A.A., Chaudhuri, P.K., Tan, D.S., Lim, W.T., Lee, S.C., Chen, P.C., Lim, C.T., Han, J., "Slanted spiral microfluidics for the ultra-fast, label-free isolation of circulating tumor cells," *Lab on a Chip* 2014; **14**(1): 128–137.

4 Movassaghi, S., Abolhasan, M., Lipman, J., Smith, D., Jamalipour, A. "Wireless body area network: a survey," *IEEE Communications Surveys and Tutorials* 2014; **16**(3): 1658–1686.

5 Maloberti, F., Davies, A.C. (eds) (2016) *"A Short History of Circuits and Systems,"* River Publishers, Delft/Aalborg.

6 Saltzman, W.M. (2009) *"Biomedical Engineering—Bridging Medicine and Technology,"* Cambridge University Press, Cambridge, NY.

7 Sannino, G., "All-on-4 concept: a 3-dimensional finite element analysis," *The Journal of Oral Implantology* 2015; **41**(2): 163–171.

8 Sannino, G., Cianca, E., Hamitouche, C., Ruggieri, M., "M2M communications for intraoral sensors: a wireless communications perspective." In: *Future Access Enablers for Ubiquitous and Intelligent Infrastructures: First International Conference, FABULOUS 2015*, Institute for Computer Sciences, Social Informatics and Telecommunications Engineering (eds), Lecture Notes of the Institute for Computer Sciences, Social Informatics and Telecommunications Engineering, vol. **159**, Springer, Ohrid, Republic of Macedonia, pp. 118–124, 2015.

9 Ruggieri, M., Prasad, R., De Sanctis, M., Rossi, T., "Integration of communications, navigation, sensing and services for quality of life: challenges, design and perspectives. In: *Communications, Navigation, Sensing and Services (CONASENSE)*, Ligthart, L. P., Prasad, R. (eds), River Publishers, Aalborg, 2013.

10 Ruggieri, M., "Telecommunications." *IX Appendix* of the Treccani Italian Encyclopedia, Treccani, Italy, 2016.

11 Prasad, R., "Human-bond wireless communications," Wireless World Research Forum, May 20–22, 2014, Marrakech, Morocco.

12 Prasad, R., "Human bond communications," *Wireless Personal Communications* 2016; **87**(3): 619–627.

13 Del Re, E., Morosi, S., Mucchi, L., Ronga, L.S., Jayousi, S., "Future wireless systems for human bond communications," *Wireless Personal Communications* 2016; **88**(1): 39–52.

14 Proceedings of First IEEE Conference on Network "Softwarization" NetSoft, London, April 2015.

15 Kim, H. and Feamster, N., "Improving network management with software defined networking," *IEEE Communication Magazine* 2013; **51**(2): 114–119.

16 Sannino, G., Barlattani, A.. "Straight versus angulated abutments on tilted implants in immediate fixed rehabilitation of the edentulous mandible: a 3-year retrospective comparative study," *The International Journal of Prosthodontics* 2016; **29**(3): 219–226.

17 Peng, Q., Budinger, T.F., "ZigBee-based wireless intra-oral control system for quadriplegic patients," *Conference Proceedings: Annual International Conference of the IEEE Engineering in Medicine and Biology Society* 2007; **2007**: 1647–1650.

18 Sardini, E., Serpelloni, M., Fiorentini, R., "Wireless intraoral sensor for the physiological monitoring of tongue pressure," 2013 Transducers & Eurosensors XXVII: The 17th International Conference on Solid-State Sensors, Actuators and Microsystems (TRANSDUCERS & EUROSENSORS XXVII), IEEE, Barcelona, Spain, June 16–20, 2013, pp. 1282–1285.

19 Park, H., Kim, J., Ghovanloo, M., "Development and preliminary evaluation of an intraoral tongue drive system," *Conference Proceedings: Annual International Conference of the IEEE Engineering in Medicine and Biology Society* 2012; **2012**: 1157–1160.

20 Park, H., Kiani, M., Lee, H., Kim, J., Block, J., Gosselin, B., Ghovanloo, M., "A wireless magnetoresistive sensing system for an intraoral tongue-computer interface," *IEEE Transactions on Biomedical Circuits and Systems* 2012; **6**(6): 571–585.

21 Ro, J.H., Kim, I.C., Jung, J.H., Jeon, A.Y., Yoon, S.H., Son, J.M., Ye, S.Y., Jeon, G.R., "System development of indwelling wireless pH telemetry of intraoral acidity," Sixth International Special Topic Conference on Information Technology Applications in Biomedicine, 2007 (ITAB 2007), IEEE, November 8–11, 2007, pp. 302–305.

22 Cho, N., Yoo, J., Song, S., Lee, J., Jeon, S., Yoo, H. "The human body characteristics as a signal transmission medium for intra-body communication," *IEEE Transactions on Microwave Theory and Techniques* 2007; **55**(5), 1080–1086.

23 Delnavaz, A., Voix, J. "Flexible piezoelectric energy harvesting from jaw movements," *Smart Materials and Structures* 2014 **23**: 1–8.

24 Mannoor, M.S., Tao, H., Clayton, J.D., Sengupta, A., Kaplan, D.L., Naik, R.R., Verma, N., Omenetto, F.G., McAlpine, M.C. "Graphene-based wireless bacteria detection on tooth enamel," *Nature Communications* 2012; **3**: 763.

25 Diaz Lantada, A., González Bris, C., Lafont Morgado, P., Sanz Maudes, J. "Novel system for bite-force sensing and monitoring based on magnetic near field communication," *Sensors* 2012; **12**: 11544–11558.

26 Sannino, G., Cianca, E., Coletta, M., Prasad, R., Ruggieri, M., Sbardella, D. "Integrated wireless and sensing technology for dentistry: an early warning system for implant-supported prosthesis," 2015 IEEE International Symposium on Systems Engineering (ISSE), Rome, September 28–30, 2015.

7

Human Bond Communication Using Cognitive Radio Approach for Efficient Spectrum Utilization

Sachin Sharma and Seshadri Mohan

Systems Engineering, University of Arkansas at Little Rock, Little Rock, AR, USA

7.1 Introduction

HBC may be defined as communication between human beings involving not only visual and auditory senses but also through other human senses: smell, taste, and touch. The information communicated may impact multiple aspects of human life—social, personal, and professional—which can be examined from theoretical and practical viewpoints. The approach opens up challenging opportunities for researchers, application developers, and service providers to improve user experience. The techniques for HBC from different perspectives may be compared and contrasted. Spurred on by the exponential growth in the usage and applications of wireless as well as connected wireless devices/sensors in homes, enterprises, vehicles, and the environment, the demand for spectrum has grown tremendously. It is reasonable to assume that HBC may bring about further demand for spectrum. Radio spectrum is a scarce natural resource that must be utilized efficiently and prudently. Federal Communications Commission studies [1] have, however, shown that spectrum is only partially and inefficiently used most of the time during a day and in many areas around the country. A well-known technique to facilitate spectrum sharing and improve spectrum utilization is by the use of cognitive radio (CR), originally proposed by Mitola [2], that is capable of learning about, and adapting to, the environment. While much work has been done in the areas of spectrum sensing, channel estimation, cooperative spectrum sensing, and cognitive relay networking, more work is required into understanding the nature of the cognitive cycles at the network, session, and application level and what should the policies be that would facilitate efficient utilization of the spectrum. In Ref. [3],

Human Bond Communication: The Holy Grail of Holistic Communication and Immersive Experience, First Edition. Edited by Sudhir Dixit and Ramjee Prasad.
© 2017 John Wiley & Sons, Inc. Published 2017 by John Wiley & Sons, Inc.

Haykin proposes a three-state cognitive cycle for radio environment sensing and spectrum management, which he terms "brain-empowered wireless communications." What we propose in this paper is an integrated approach that encompasses all levels of the protocol stack of a system with a similar definition and approach to the cognitive cycle at all the other layers of the system as defined in Ref. [3] at the radio environment. Also any approach we propose needs to account the CR end-user behavior, bandwidth needs, quality of service (QoS) requirements, policies for spectrum sharing, security concerns, and other such requirements. This paper aims to address these needs by formulating a cognitive cycle framework at the systems level. Alternatively, this paper seeks answer to the question "How can the entire CR system at all the protocol layers be brain-empowered?" This chapter identifies machine learning (ML) as an inherent approach and an integral tool for the CR framework at the systems level and aims to explore the synergy between ML and CR that could lead to potentially transformative research and provide new insights into the analysis and innovative ways to build ML-based CR systems that seek to optimize spectrum utilization through the formulation of novel and transformative strategies for spectrum sharing.

HBC-based social applications may be proposed that could serve to enhance the experience of spectators in a sports arena or to facilitate the passengers in vehicles forming a vehicular ad hoc network (VANET) to share information about the traffic and road conditions that could lead to efficient utilization of the transport infrastructure and reduce traffic congestion. The ML-based CR theories and algorithms developed will lead to spectrally efficient algorithms for social networking and improve spectrum utilization. We anticipate that this research will point to the feasibility of a new concept, that of cognitive radio network service provider (CRNSP) who may not have an originally assigned spectrum similar to the concept of mobile virtual network operator (MVNO) without original infrastructure but leases from wireless service providers (WSP).

Since the introduction of CR by Mitola [2], researchers have carried out extensive studies that focus on spectrum sensing techniques, that is, the problem of detecting the unoccupied channels in the licensed frequency bands. Different algorithms with different scenarios have been developed for spectrum sensing [4–20] that focuses on minimizing detection errors and sensing delay. The algorithms are designed to reduce the probability of interrupting the primary system and increase the time available for data transmission. Overall, the CR system throughput will be maximized by achieving these objectives. Reference [21] provides a survey of different types of setups, challenges, and detection methods for spectrum sensing. In Ref. [22], an optimal sensing framework has been developed to maximize the opportunistic spectrum access efficiency while taking into consideration the interference constraints for a single band. In addition, a resource allocation procedure is developed to maximize the spectrum access opportunities.

Game theory provides an approach to decision-making in various complex scenarios. The theory offers two types of models: cooperative and noncooperative. Cooperative game theory is a technique that involves a cooperative game playing environment among multiple players to strike coalitions among them, thereby impacting their strategies dynamically. This technique focuses on the benefits of the coalition among multiple players.

Cooperative game theory [23] provides a tool for such algorithms where collaborative spectrum sensing problem can be modeled among the set of players. Noncooperative game theory offers a technique that allows for a competitive game playing environment where every player makes a decision in their favor independently. Each player tries to optimize their gains or their individual preferences. In Ref. [3] a game-theoretic decision-making tool plays a key role in the design of a CR system. Reference [24] formulates an adaptive channel allocation scheme to achieve an optimal solution for cognitive radio networks (CRN) using game theory. In Ref. [25], Nash equilibrium-based game is described, where no player can increase the payoff by choosing a different action. In Ref. [26], Nelder–Mead direct search method is used to achieve the Nash equilibrium. To achieve such equilibrium state and network performance of transmitting nodes in wireless networks, game models with multiple solutions have been proposed [27]. An evolutionary game model has been introduced in Ref. [28], where players interact with other players and modify their strategies in a dynamic way, formulated for ALOHA protocol. In Ref. [29] a medium access contention game model is proposed for CSMA protocol for successful data transmission. In Ref. [30] a noncooperative game model is proposed in a point-to-multipoint network for best traffic network management using contention control. In Ref. [31], a cooperative game model is proposed for wireless mesh network architecture to communicate among routers and clients. In Ref. [32], the authors discuss a computational model of energy consumption to optimize the problem in a dynamic scenario using game theory. In Ref. [33], the game theory has been applied into a multidisciplinary design optimization problem in a noncooperative environment that is based on gene expression programming. Our work in Ref. [34] involves the placement of access points (APs) among users in an enterprise so as to distribute the users among APs to optimize the network throughput.

One can envision that HBC will take on added importance in a VANET scenario. HBC applications may facilitate passengers in vehicles to inform each other of ongoing local events and share their experiences. In Ref. [35] the authors propose an approach to provide a broader coverage with many applications for communication among vehicles with reduced latency to users. Continuous connectivity in the vehicular mobility scenario is a challenge to address and incorporate within the infrastructure of VANETs. In Ref. [36], the lifetime of different communication links among multiple nodes has been discussed. In Ref. [37], such links are analyzed via simulation to find the best solution in the highway scenario. Another challenge is safety communication among vehicles in both safe and unsafe situations. In Ref. [38], the different safety communication examples have been discussed using various parameters and congestion problem has been raised as well. The objective of vehicle safety with reduced power consumption provides the platform to the various researches in VANETs. Aspects that impart an intelligent platform to vehicles using various embedded sensors that facilitate passengers to engage in safe HBC-different security architectures, security concepts, and security issues are needed. In Ref. [39], the authors proposed a technique for data delivery among vehicle-to-vehicle (V2V) communication under various traffic patterns and infrastructure layout. In Ref. [40], the authors proposed an estimation approach to find a relationship between network density and various traffic flows under different scenarios. The proposed estimation facilitates the vehicles to alter the range of transmission and possibly increase network throughput and performance. In Ref. [41], the authors describe a protocol for the discovery of Internet gateways (IGW) that would facilitate the vehicles to access Internet services in future intelligent vehicle communication systems. An important constraint in intelligent communication among multiple vehicles in VANET is security, which is a considerable challenge, and such communication must be carried out securely, especially when intermediate vehicles are used by a vehicle to communicate with an IGW. Different security protocols have been proposed in a VANET environment to assign an electronic unique certificate id [42], group signature and identity-based signature certificate [43], and digital signature [44] to an individual vehicle. In Ref. [45], an evaluation algorithm has been proposed that performs the spectrum sensing and decision-making process regarding unused frequency bands and methods to aggregate individual sensing data.

7.2 Human Bond Communication Using Cognitive Radio Approach for Efficient Spectrum Utilization

Since this chapter focuses on CR systems for HBC, it is useful to consider a framework for the cognition cycle model for the entire CR system covering the various protocol layers. Independently, researchers have worked on ML and developed various models for ML. It is essential to understand the use of ML tools and techniques and their applicability to the process of cognition. The research here suggests the use of ML-based approaches to CR systems. We anticipate that such interdisciplinary approaches combining ML and CR systems could lead to new approaches that could be applied to serve the needs of HBC, or of a community such as spectators in a sports arena, to VANETs to a neighborhood, each sharing some common interests. It is therefore essential to focus on developing dynamic spectrum sharing methodology to improve spectrum band utilization and performance. This leads us to consider the various aspects of communications required for local user satisfaction and spectrum management among them in a dynamic scenario. The learning capabilities from the present and past observations helps a CR system equipped with ML systems to become aware of spectrum availability, user demands and QoS needs, and network and radio environment characteristics. An ML-based CR system will be capable of achieving multifaceted capabilities including perception, learning, and reasoning along with self-classification and self-organization capabilities. Learning process imparts to a CR system the ability to collect the required knowledge or data and, from the collected data, extract various parameters from the various environments [46]. ML algorithms have been proposed in CRN that can be categorized into two types: supervised learning and unsupervised learning. In these learning processes, CR makes the decision after the observation of the current state. The current state consists of real-time environment. The history of the real-time environment plays a key role in the learning and optimization process [47]. In CRN, when multiple agents try to observe the different scenarios and optimize their process simultaneously, techniques must be devised that optimize the overall utilization of the spectrum. By applying cooperative or noncooperative game scenarios, learning capabilities and spectrum utilization can be optimized.

According to Ref. [48], in CR, Q-learning may provide a suitable framework to estimate the actions of the network nodes. Q-learning is one type of reinforcement learning technique. An appropriate training dataset is required to optimize the performance of spectrum sensing process, spectrum management, and decision-making policy process. Reinforcement learning allows CR to modify its behavior by interacting with various environments under certain or uncertain conditions [49]. It optimizes the various processes by adjusting several parameters associated with the processes. According to Ref. [50], in a centralized system, a real-time learning algorithm can be used to obtain optimal solution, without having the prior knowledge of the transition probabilities. Reinforcement learning will obtain the information for a CR about the upcoming available spectrum channel band at a particular geographical location using rewards, actions, and states. This prior information about the spectrum channel band will help to reduce the scanning time and power consumption by CR. Using this learning method CR will observe the different parameters, create the different states, and action plans. After analyzing those action plans based on the rewards, the CR system may optimize different parameters. A deep learning method that has capabilities to ensure the accuracy of domains like speech recognition was observed [51]. In Ref. [52], deep learning algorithm that has the capability to learn complex concepts with extremely supervision can be used in spectrum sensing and spectrum management. According to Ref. [52], in a variety of applications, performance can be improved by incorporating deep learning technique into a CR system.

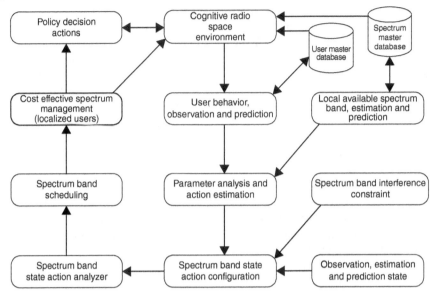

Figure 7.1 Cognition cycle for the CR system.

Figure 7.1 illustrates an enhanced cognition cycle for the CR system that takes into account, in addition to the CR environment and spectrum availability as provided by the Spectrum Master database, the behavior of users concerning spectrum usage, their observation regarding spectrum sensing and estimation of the quality of channels, and other measurements they may make and wish to share, which are then made available in a user master database. The user behavior that impacts the cognition cycle may be different at different phases. For example, user behavior may refer to their willingness to forget spectrum usage during certain hours of the day. The user of WSP may then collect such information and lease out spectrum to a CRNSP for use by CR users as will be further explained later. From the data collected, using ML, useful data can be extracted that will help in the allocation of spectrum bands to users taking into account their requirements, for example, QoS needs, so as to optimize spectrum utilization.

Since the motivation behind this idea is to strike a synergy between traditional CR approaches and ML and take into account in our models user behavior and requirements, we suggest that this interdisciplinary approach take into account the following:

1) While allocating channels to CR users, ensure adequate QoS is provisioned so as to meet their expected or specified requirements such as bandwidth, delay, and jitter.

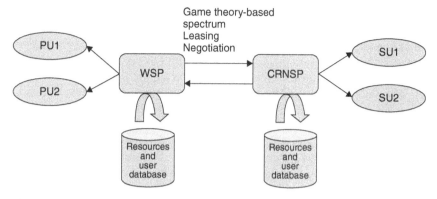

Figure 7.2 Dynamic spectrum leasing methodology/model (DSLM).

2) Ensure traffic management so as to avoid saturation of spectrum.
3) Apply reinforcement learning techniques by defining suitable states of users (based on their behavior of spectrum usage) and reward functions to users based on their actions to optimize channel sensing, channel allocation, and/or channel utilization, which essentially form the tasks of spectrum management.
4) Invoke deep learning techniques to help understand the users' behavior and ensure that channels are allocated to users so as to meet the users' QoS needs in real time in an adaptive manner.
5) Ensure that application level cognition model we will evolve facilitates application level networking or social networking in a V2V scenario and will spur the development of useful applications of CRN technology that will be attractive to users.

As an example of the research methodology adopted, we propose a game theory-based learning approach to dynamic spectrum leasing. We propose a scenario in which a traditional WSP and a CRNSP introduced earlier interact with each other using a cooperative game playing and learning strategy to share spectrum. In Figure 7.2, PU denotes a primary user of WSP and SU denotes a secondary user of CRNSP. The figure depicts the interaction between WSP and CRNSP that will facilitate CRNSP to lease spectrum from WSP.

Spectrum leasing is one of the approaches that has been suggested, an alternative to the exclusive-use model, in which the spectrum licensees are granted the rights to sell or trade their spectrum to third parties [5, 6]. However, spectrum leasing has been identified as a static, or offline, spectrum sharing technique, with the possible exception as suggested in Ref. [7]. However, the spectrum leasing techniques proposed in Ref. [7] differ from our DSLM approach in several ways. The low utilization of the licensed spectrum remains a primary concern, and the wasted resources can possibly be used in communication for high bandwidth and low latency requirements [8]. Reference [9] provides an interesting survey and suggests possible solutions for CRN. Very little work has been done to support traffic with QoS in CRNs. For example, the performance for transmitting voice traffic in a CRN is studied in Refs. [10, 11], where a single channel is shared by the CRN and the primary network. Energy efficiency in a CRN with multi-carrier modulation is studied in Refs. [12, 13]. In Ref. [14] the performance of a CRN for supporting communication traffic is studied. In Ref. [15], a game-theoretic formulation is provided for dynamic spectrum sharing between PUs and SUs. We propose to formalize the concept of developing a game-theoretic framework for dynamic spectrum sharing and articulate the DSLM model. The approach proposed here will facilitate the use of HBC applications by VANET users.

In our proposed DSLM, similar to Ref. [15], we consider four sets of users. While in Ref. [15], the PUs and SUs directly interact with each other in sharing the spectrum; in our model, they do not. Instead, the PUs may choose to give up their use of the licensed spectrum during certain hours and may inform their WSP. In return, the WSP rewards their users for their actions. The WSP dynamically updates the spectrum bands that are available and leases the available spectrum to CRNSP. Thus, in our model WSP and CRNSP interact with each other to dynamically lease the spectrum and, thus unlike in Ref. [15], introduce a layer of security between PUs and SUs. The four sets of users are defined as follows:

- Primary User I (PU1): User does not opt for the reward program offered by WSP.
- Primary User II (PU2): User who is willing to opt for the reward program of WSP and relinquish the use of spectrum during non-busy hours to the WSP.
- Secondary User I (SU1): User who subscribes to the CR services offered by CRNSP.
- Secondary User II (SU2): User who is a default user and has not subscribed to any services from CRNSP but can utilize the services as needed.

Such a model lends itself to game-theoretic analysis. In Figure 7.1, we propose a cooperative game theory approach to be followed by WSP and CRNSP. Using this approach both service providers can work together on leasing agreement to raise their revenue and fulfill their user requirements. WSP can share the information of the unused available spectrum bands of the licensed users with CRNSP. By sharing this information with CRNSP, WSP can increase its revenue. Using this shared information CRNSP can satisfy their user's requirements. We propose noncooperative game theory approach to be followed by WSP and CRNSP to satisfy their users' requirement independently of each other. We have categorized WSP users into two types, PU1 and PU2. PU1 is not interested in enrolling in the reward program, while PU2 enrolls in the reward program offered by WSP to reduce the cost of subscribing to services from WSP.

Periodically, WSP and CRNSP engage in a cooperative game-theoretic sense to compute a utility function and determine their strategies until the game converges. The utility function is defined as revenue minus cost. For the WSP, the spectrum leased out brings forth revenue from CRNSPs, and the cost would be the loss of spectrum for its own users and interference that the use of leased spectrum by CR users may bring about. For the CRNSP, the revenue is derived through CR users subscribing to CRNSP services or using the CR services in an opportunistic manner and the cost arises from leasing the spectrum from WSP.

Our preliminary study of CRNSP and CR users' gaming strategy and learning leads to the result shown in Figure 7.3. The details are however omitted due to lack of space, but a formulation of a game-theoretic strategy for the placement of femtocell APs appears in Ref. [34]. It is interesting to observe that the utility value increases with an increasing number of spectrum bands, that is, with increasing cooperation between WSP and CRNSP. CRNSP receives more spectrum bands from WSP, which are then offered to CR users to derive increased revenue that increases the utility function.

With another example of the research methodology that we will employ in this study, we briefly describe the application of ML to spectrum management.

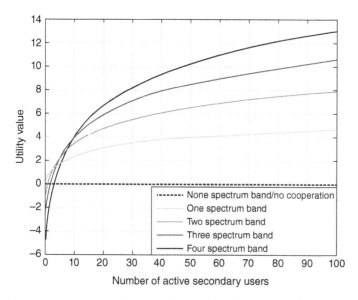

Figure 7.3 Utility function versus the number of active secondary users, parameterized by the number of spectrum bands.

What is missing today are systematic studies to understand how advances in deep learning can be applied to inference tasks relevant to spectrum sensing in real time. We envision that the synergy between experts in CR and ML will catalyze competitive future research proposals on efficient sensing management strategies guided by data mining and optimization techniques. The idea is to develop a tool for use by WSP and CRNSP for real-time and offline spectrum management. The tool learns from the data uploaded to a user master database (Figure 7.1) using deep learning and reinforced learning techniques and performs assignment of spectrum to CR users that maximizes utilization of spectrum dynamically acquired by CRNSP through the game-theoretic strategy as described previously. Further research is needed to explore the application of deep learning techniques in CR, spectrum sensing, and management. Based on the requirement of the user, this technique will facilitate the allocation of appropriate spectrum to each CRNSP user. Spectrum sensing, noise interference, channel utilization, energy-efficient transmission, channel noise, path loss, transmit power, channel throughput, QoS, and reliable communication among end users may be optimized using this multidisciplinary approach.

Data collection and data mining techniques will assume paramount significance.

We propose a framework for the multidisciplinary study as in Figure 7.4, depicting a seven-phased architecture and model of a cognition cycle. Each phase consists of several subtasks—(i) collecting data; (ii) configuring the data and formulating or choosing techniques applicable at each phase for the analysis of collected data; (iii) formulating, validating, and improving the cognition cycle models; and (iv) formulating the final model as output of each phase. The seven phases are briefly explained in the following:

1) Spectrum sensing and channel estimation techniques: This phase of the research involves implementing many of the techniques developed in the literature.
    - Input: Channel sensing and estimation algorithms and users' behavior regarding usage of channels.
    - Output: Formulation and analysis of a cognitive cycle model that incorporates and adapts sensing and estimation strategies so as to minimize battery power consumption while meeting user requirements.
2) Spectrum management using ML at the network layer: Data is collected about the behavior of users in a VANET setting concerning the use of the spectrum, their requirements, and spectrum availability in a dynamic manner. Available channels are matched and allocated to the users dynamically. ML (deep learning and reinforced learning) techniques are applied to the process of spectrum management to maximize efficient spectrum utilization.
    - Input: Users' behavior regarding usage of channels.
    - Output: A cognition cycle model at the network layer, new algorithms and their analyses, and new contributions to the fields of CRN and ML.

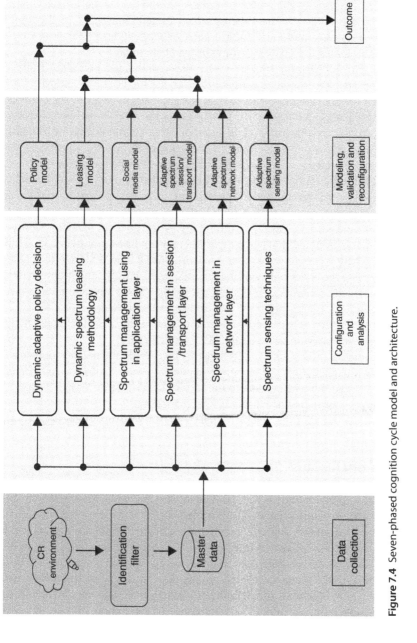

Figure 7.4 Seven-phased cognition cycle model and architecture.

3) Spectrum management using ML at the session layer: At the session layer, the issues that are of concern include maintaining an ongoing session between two vehicular users independent of their relative speeds. Techniques from ML may be applied to determine channel resources for the session and ensure that backup channels are available should the currently used channels for the session do not meet the user needs due to low signal quality due to, for example, channel noise, interference, or shadowing.
   - Input: Users' behavior regarding usage of channels.
   - Output: A cognition cycle model at the session layer, new algorithms and their analyses, and new contributions to the field of CRN and ML.
4) Spectrum management using ML at the application layer: In the context of a VANET and a V2V scenario, this phase of the work involves collection of blog data involving traffic conditions or other items of interest to passengers involved and formulation of a cognition cycle at the application layer.
   - Input: Social user behavior, user requirements, and blog data.
   - Output: A cognition cycle model at the application layer and formulation, analysis, validation, and improvement of novel spectrum sharing and assignment algorithms for the VANET V2V scenario.
5) Dynamic spectrum leasing methodology: Novel concept of a CRNSP is proposed. It is assumed that while a CRNSP may not have an assigned spectrum, the CRNSP is able to lease spectrum from a WSP. The WSP is in turn able to determine the users' behavior regarding spectrum usage and is able to determine what spectrum can be leased out to the CRNSP on a dynamic basis.
   - Input: WSP users' behavior regarding spectrum usage.
   - Output: A dynamic spectrum leasing algorithm that facilitates efficient spectrum sharing between the WSP and CRNSP.
6) Dynamic adaptive policy decision-making: WSP formulates policies for bandwidth allocation and resources needed to ensure QoS so as to maximize user satisfaction with respect to the service needs (Figure 7.4).
   - Input: User behavior regarding spectrum usage.
   - Output: An adaptive policy model.

7.3 Conclusions

A multidisciplinary study is proposed involving ML and CR for efficient spectrum usage. A framework is proposed based on which this study will likely evolve in the future. A dynamic spectrum leasing methodology is proposed for sharing spectrum between a WSP and a CRNSP. The spectrum acquired by CRNSP is used by its users for a number of application scenarios, including HBC and VANET scenarios. This framework proposes collecting, correlating, and analyzing thousands of attributes, including real-time user behaviors and available spectrum situational information. ML algorithms may be employed to provide intelligence evaluation and spectrum predictions and ensure overall user satisfaction based on the gathered information. Cooperative and noncooperative game-theoretic strategies could be applied toward optimizing appropriate revenue function. The required operational changes are automatically incorporated into the system's datasets. Detection of advanced security threat using user analytics may be implemented through ML and modeling techniques to incorporate information about users and their behavior including their requirements, access time, activity type, geographic location data gathered from network infrastructure, and alerts from security solutions. This analysis data is correlated and analyzed based on past and real-time activities of a CR user. Some analysis like session duration, resources used, connectivity status, and group behavior may also be taken into account.

Some thoughts on future HBC scenarios are as follows:

- Virtual reality on your cell phone
- Cell phone as magical wand for storage of human thoughts, memory
- Application of nano, bio, info sensors for HBC
- Virtual presence at a remote place with real HBC with real and virtual conference participants
- HBC utilizing nano and quantum communications and networks

References

1 Federal Communications Commission, "Spectrum Policy Task Force," Report of the Spectrum Efficiency Working Group ET Docket no. 02-135, Nov. 2002.

2 J. Mitola et al., "Cognitive radio: making software radios more personal," *IEEE Personal Communications*, vol. **6**, no. 4, pp. 13–18, Aug. 1999.

3 S. Haykin, "Cognitive radio: brain-empowered wireless communications," *IEEE Journal on Selected Areas in Communications*, vol. **23**, no. 2, pp. 201–220, Feb. 2005.

4 W. Zhang and K. B. Letaief, "Cooperative spectrum sensing with transmit and relay diversity in cognitive radio networks," *IEEE Transactions on Wireless Communications*, vol. **7**, no. 12, pp. 4761–4766, Dec. 2008.

5 W. Zhang, R. K. Mallik, and K. B. Letaief, "Optimization of cooperative spectrum sensing with energy detection in cognitive radio networks," *IEEE Transactions on Wireless Communications*, vol. **8**, no. 12, pp. 5761–5766, Dec. 2009.

6 D. Duan, L. Yang, and J. C. Principe, "Cooperative diversity of spectrum sensing for cognitive radio systems," *IEEE Transactions on Signal Processing*, vol. **58**, no. 6, pp. 3218–3227, Jun. 2010.

7 Z. Quan, S. Cui, and A. H. Sayed, "Optimal linear cooperation for spectrum sensing in cognitive radio networks," *IEEE Journal of Selected Topics in Signal Processing*, vol. **2**, no. 1, pp. 28–40, Feb. 2008.

8 Y. Zeng, and Y. Liang, "Maximum–minimum eigenvalue detection for cognitive radio," In *Proceedings of the IEEE 18th International Symposium on Personal, Indoor, and Mobile Radio Communication (PIMRC)*, Athens, Greece, Sep. 2007.

9 T. H. Lim et al., "GLRT-based spectrum sensing for cognitive radio," In *Proceedings of the IEEE Global Telecommunications Conference (GLOBECOM)*, New Orleans, LA, pp. 1–5, Nov. 2008.

10 D. Cabric, S. M. Mishra, and R. W. Brodersen, "Implementation issues in spectrum sensing for cognitive radios," In *Proceedings of the Asilomar Conference on Signals, Systems, and Computers*, Pacific Grove, CA, pp. 772–776, Nov. 2004.

11 H. Tang, "Some physical layer issues of wide-band cognitive radio systems," In *Proceedings of the IEEE International Symposium on New Frontiers in Dynamic Spectrum Access Networks*, Baltimore, MD, pp. 151–159, Nov. 2005.

12 S. Shankar, C. Cordeiro, and K. Challapali, "Spectrum agile radios: utilization and sensing architectures," In *Proceedings of the IEEE International Symposium on New Frontiers in Dynamic Spectrum Access Networks*, Baltimore, MD, pp. 160–169, Nov. 2005.

13 V. K. J. Lunden, A. Huttunen, and H. V. Poor, "Multiple sensing in cognitive radios based on multiple cyclic frequencies," In *Proceedings of the International Conference on Cognitive Radio Oriented Wireless. Networks and Communications*, Orlando, FL, pp. 1583–1590, Aug. 2007.

14 K. E. N. P. D. Sutton, and L. E. Doyle, "Cyclostationary signatures in practical cognitive radio applications," *IEEE Journal on Selected Areas in Communications*, vol. **2**, no. 1, pp. 13–24, Jan. 2008.

15 Z. Tian, and G. B. Giannakis, "A wavelet approach to wideband spectrum sensing for cognitive radios," In *Proceedings of the International Conference on Cognitive Radio Oriented Wireless. Networks and Communications*, Mykonos Island, Greece, pp. 1–5, Jun. 2006.

16 A. Ghasemi, and E. S. Sousa, "Collaborative spectrum sensing in cognitive radio networks," In *Proceedings of the IEEE International Symposium on New Frontiers Dynamic Spectrum Access Networks*, Baltimore, MD, pp. 131–136, Nov. 2005.

17 G. Ganesan, and Y. G. Li, "Cooperative spectrum sensing in cognitive radio, part I: two user networks," *IEEE Transactions on Wireless Communications*, vol. **6**, no. 6, pp. 2204–2213, Jun. 2007.

18 G. Ganesan, and Li Ye, "Cooperative spectrum sensing in cognitive radio, part II: multiuser networks," *IEEE Transactions on Wireless Communications*, vol. **6**, no. 6, pp. 2214–2222, Jun. 2007.

19 B. B. G. Ganesan, Y. G. Li, and S. Li, "Spatiotemporal sensing in cognitive radio networks," *IEEE Journal on Selected Areas in Communications*, vol. **28**, no. 1, pp. 5–12, Jan. 2008.

20 A. S. S. M. Mishra, and R. W. Brodersen, "Cooperative sensing among cognitive radios," In *Proceedings of the IEEE International Conference on Communications*, vol. 4, Istanbul, Turkey, pp. 1658–1663, Jun. 2006.

21 T. Yucek, and H. Arslan, "A survey of spectrum sensing algorithms for cognitive radio applications," *IEEE Communications Surveys & Tutorials*, vol. 11, no. 1, pp. 116–130, Jan. 2009.

22 W.-Y. Lee, and I. Akyildiz, "Optimal spectrum sensing framework for cognitive radio networks," *IEEE Transactions on Wireless Communications*, vol. 7, no. 10, pp. 3845–3857, Oct. 2008.

23 R. B. Myerson, *Game Theory, Analysis of Conflict*. Cambridge, MA: Harvard University Press, Sep. 1991.

24 N. Nie, and C. Comaniciu, "Adaptive channel allocation spectrum etiquette for cognitive radio networks," In *Proceedings of the IEEE DySPAN'05*, Baltimore, MD, pp. 269–278, Nov. 2005.

25 M. J. Osborne, *An Introduction to Game Theory*. Oxford, UK: Oxford University Press, 2003.

26 J. A. Neldel, and R. Mead, "A simplex method for function minimisation," *The Computer Journal*, vol. 7, pp. 308–313, 1965.

27 M. Felegyhazi, and J.-P. Hubaux, "Game Theory in Wireless Networks: A Tutorial," EPFL Technical Report, LCA-REPORT-2006-002, EPFL, Lausanne, Switzerland, Feb. 2006.

28 H. Tembine, E. Altaian, and R. El-Azouzi, "Delayed evolutionary game dynamics applied to medium access control," In *Proceedings of the IEEE International Conference on Mobile Ad hoc and Sensor Systems 2007*, Pisa, Italy, pp. 1–6, Oct. 8–11, 2007.

29 Y. Cho, C. S. Hwang, and F. A. Tobagi, "Design of robust random access protocols for wireless networks using game theoretic models," In *Proceedings of the 27th IEEE International Conference on Computer Communications (IEEE INFOCOM 2008)*, Phoenix, AZ, pp. 1750–1758, Apr. 13–18, 2008.

30 S. Chowdhury et al., "Game-theoretic modeling and optimization of contention-prone medium access phase in IEEE 802.16/WiMAX networks," In *Proceedings of the Third International Conference on Broadband Communications, Information Technology and Biomedical Applications 2008*, Pretoria, South Africa, pp. 335–342, Nov. 2008.

31 L. Zhao, J. Zhang, and H. Zhang, "Using incompletely cooperative game theory in wireless mesh networks," *IEEE Network*, vol. 22, no. 1, pp. 39–44, Jan.–Feb. 2008.

32 Zhijiao Xiao et al., "A solution of dynamic VMs placement problem for energy consumption optimization based on evolutionary game theory," *Journal of Systems and Software*, vol. 101, pp. 260–272, Mar. 2015.

33 Mi Xiao et al., "A new methodology for multi-objective multidisciplinary design optimization problems based on game theory," *Expert Systems with Applications*, vol. 42, no. 3, pp. 1602–1612, Feb. 2015.

34 H. Shyllon, and S. Mohan, "A game theory based distributed power control algorithms for Femto cells," In *2014 IEEE ANTS*, New Delhi, India, Dec. 2014.

35 S. Yousefi, S. Bastani, and M. Fathy, "On the performance of safety message dissemination in vehicular ad hoc networks," In *IEEE Fourth European Conference on Universal Multiservice Networks*, European, 2007.

36 S. Y. Wang, "Predicting the lifetime of repairable unicast routing paths in vehicle-formed mobile ad hoc networks on highways," In *The 15th IEEE International Symposium on Personal, Indoor and Mobile Radio Communications (PMRC)*, Barcelona, Spain, September 5–8, 2004.

37 J. Haerri, and C. Bonnet, "A lower bound for vehicles' trajectory duration," In *Proceedings of the IEEE 62nd Semiannual Vehicular Technology Conference (VTC Fall 05)*, Dallas, TX, Sep. 2005.

38 N. Yang, J. Liu, and F. Zhao, "A vehicle-to-vehicle communication protocol for cooperative collision warning," In *First Annual International Conference on Mobile and Ubiquitous Systems (MobiQuitous'04), Networking and Services*, Cambridge, MA, August 22–25, 2004.

39 J. Zhao, and G. Cao, "VADD—vehicle-assisted data delivery in vehicular ad hoc networks," In *IEEE INFOCOM*, Barcelona, Spain, Apr. 2006.

40 M. M. Artimy, W. Robertson, and W. J. Phillips, "Assignment of dynamic transmission range based on estimation of vehicle density," In *ACM VANET'05*, Cologne, Germany, Sep. 2005.

41 M. Bechler et al., "Efficient discovery of internet gateways in future vehicular communication systems," In *The 57th IEEE Semiannual Vehicular Technology Conference*, Jeju, Korea, Apr. 22–25, 2003.

42 M. Raya, and J.-P. Hubaux, "Securing vehicular ad hoc networks," *Journal of Computer Security—Special Issue on Security of Ad-hoc and Sensor Networks*, vol. **15**, no. 1, pp. 39–68, 2007.

43 X. Lin et al., "GSIS: a secure and privacy-preserving protocol for vehicular communications," *IEEE Transactions on Vehicular Technology*, vol. **56**, no. 6, pp. 3442–3456, Nov. 2007.

44 P. S. L. M. Barreto et al., "Efficient and provably-secure identity-based signatures and signcryption from bilinear maps," In *Proceedings of the Advances in Cryptology—ASIACRYPT 2005*, Taj Coromandel, Chennai, India, pp. 515–532, Dec. 2005.

45 K. Baraka et al., "An infrastructure-aided cooperative spectrum sensing scheme for vehicular ad hoc networks," *Ad Hoc Networks*, vol. **25**, no. Part A, pp. 197–212, February 2015.

46 R. S. Michalski, "Learning and cognition," In *World Conference on the Fundamentals of Artificial Intelligence (WOCFAI'95)*, Paris, France, pp. 507–510, Jul. 1995.

47 C. Clancy et al., "Applications of machine learning to cognitive radio networks," *IEEE Wireless Communications*, vol. **14**, no. 4, pp. 47–52, Aug. 2007.

48 C. Claus, and C. Boutilier, "The dynamics of reinforcement learning in cooperative multiagent systems," In *Proceedings of the Fifteenth National Conference on Artificial Intelligence*, Madison, WI, pp. 746–752, Jul. 1998.

49 R. S. Sutton, and A. G. Barto, *Reinforcement Learning: An Introduction.* Cambridge, MA: MIT Press, 1998.

50 L. Deng et al., "Recent advances in deep learning for speech research at Microsoft," In *IEEE International Conference on Acoustics, Speech and Signal Processing, ICASSP 2013*, Vancouver, BC, Canada, May 26–31, 2013.

51 Q. V. Le et al., "Building high-level features using large scale unsupervised learning," In *29th International Conference on Machine Learning (ICML 2012)*, Edinburgh, Scotland, June 26–July 1, 2012.

52 L. Deng, and D. Yu, *Deep Learning: Methods and Applications.* Delft, the Netherlands: Now Publishers, Inc., Jan. 2014.

8

Technology Advancement and Integration in the Context of Wildlife Conservation

Pradeep K. Mathur[1], Bilal Habib[1], and Prateek Mathur[2]

[1] *Wildlife Institute of India, Dehradun, India*
[2] *CTIF, Aalborg University, Aalborg, Denmark*

8.1   Introduction

The living planet—the Earth—is endowed with extensive wilderness and enormous diversity of natural ecosystems and associated wild flora and faunal species. Wide-ranging natural ecosystems include the Arctic and Antarctic regions as north and south poles, ice-capped high mountain ranges, climatically variant cold and hot deserts, biodiversity-rich temperate and tropical forests, grasslands and savannahs, varied wetlands including river ecosystems and their floodplains, long coastlines, deep marine areas, and open oceans. They provide innumerable goods and wide-ranging ecological services besides their strong linkages with human well-being by way of direct contributions to life support system and all-round development. In spite of natural ecosystems and wildlife, resources are vital for humanity; most natural ecosystems, wilderness, and wild animal species in particular are in decline across the world. Various timely initiatives and concerted conservation efforts by the governmental and nongovernmental agencies to contain the decline are underway; still a large number of species are on the brink of extinction. This is evident based on the agenda adopted by the multilateral conservation program, the Convention on Biological Diversity (CBD), through its Aichi Biodiversity Targets for 2020. One of the key objectives is to conserve 17% of terrestrial and inland water and 10% of coastal and marine areas, especially areas of particular importance for biodiversity and ecosystem services, through the establishment of effective protected area (PA) systems [1]. Certain regions on the planet have been classified as biodiversity hotspots as they are excessively biodiversity rich. The hotspots account for as much as 35% of mammals, reptiles, birds, and

Human Bond Communication: The Holy Grail of Holistic Communication and Immersive Experience, First Edition. Edited by Sudhir Dixit and Ramjee Prasad.

amphibians that are endemic to these regions. Interestingly, hotspots account for a minuscule 2.3% of the Earth's land surface [2]. The hotspots are severely threatened due to human-induced activities and interference for deriving diverse economic utilities from these regions. The problem is compounded by the fact that majority of the hotspots are located in the tropical region, mostly within the developing countries. Collectively, these aspects demonstrate the critical and complex relationships between biodiversity conservation today and human involvement. The likelihood of attaining an increase in the PAs to attain the Aichi Biodiversity Target is a daunting task due to the lack of adequate resources, and even in several countries, much desired political and community support is lacking [3].

Natural wildlife habitats are being set aside and protected, referred to as PAs, and such conservation approach is termed as *in situ* conservation. The alternative conservation approach *ex situ* seeks protection and management of zoological parks and botanical gardens. Though numerous initiatives have been taken in the form of captive bred conservation efforts at *ex situ* sites, they are not a possible alternative to conservation efforts in PAs except in some instances can support *in situ* conservation. Further, the *ex situ* conservation measures are not easy to implement, and they come with their own set of operational challenges: weakening of gene pool and inbreeding, reduced lifespan, behavioral changes in animals, and so on. Throughout this chapter, the word *PAs* is used to refer to natural wildlife habitat. Many of the PAs have been specially designated as World Heritage Sites, biosphere reserves, tiger reserves, and Ramsar wetland sites by various global conservation agencies.

Technology-based wildlife conservation efforts have been at the forefront during past couple of decades, and today there are a plethora of mechanisms available for wildlife research, monitoring, and conservation. However, on the whole, various conservation strategies, initiatives, and technological solutions have mainly focused on the conservation of umbrella and flagship species through the use of devices in the form of radio collars, electronic tags, and camera traps. Umbrella species/charismatic species often refers to large mammals, for example, tiger, lion, leopard, elephant, and rhinoceros. Such species usually have an extensive home range, cover diverse habitats, ultimately occupy large areas for their various biological needs, and are able to elicit public support. Conservation of varied habitats in a large home range is expected to help other lesser-known and smaller species by default as they dovetail the effort and get full benefit. In some instances, successful conservation of flagship species in turn ensures landscape connectivity through wildlife corridors between two PAs, which is very significant [4]. Technological solutions have assisted in the conservation of landscapes by preventing further degradation and averting extinction threats to some of the critically endangered species. Technological solutions chiefly available in the form of information and communications technology (ICT) have been applied for wildlife research and monitoring activities, but at the same time, enormous potential of use of biotechnology such as deoxyribonucleic acid (DNA)-based technologies and genomics have been recently recognized particularly in the wildlife conservation sector. These DNA-based technologies have been specifically utilized in wildlife forensics and for protecting the gene pool of endangered species. However, technological solutions for wildlife conservation impose their own set of problems that are either constraining or influencing the conservation efforts feasible today or in the time to come. Mechanisms that can overcome these potential challenges in a comprehensive manner would be beneficial for the agenda of coexistence of wildlife and human alongside all-round rapid development that is also vital for human well-being.

The technology-based interventions and related activities toward wildlife conservation have chiefly relied on operational concepts similar to the human communication systems, wherein instead of a holistic picture, only a partial information is communicated relying on seeing and auditory sensory information. This chapter details out in the subsequent sections the progress made so far in technological advancement and integration of various interventions with wildlife conservation, the challenges imposed by such interventions, and the possibilities to address them in the near future. Further, the necessity for strong linkages between technological interventions and empowerment of conservationists and field practitioners is highlighted.

8.2   Technology-Based Wildlife Conservation

Studying wild animals in their natural habitats using traditional means has always been a daunting task. However, lately available technological solutions have greatly reduced the reliance on human-based research and monitoring of a variety of terrestrial, aquatic, and wild marine animals. The most focused aspect of the technology use in the field of wildlife research and conservation efforts has been the capacity to track the movement and dispersal of wild animals and determine their locations even in the remote locations/interiors of forests or in deep water and other inaccessible, difficult sites. This is based on the implicit relation, that is, an animal tracked periodically indirectly represents its well-being. With the advent of telemetry, studies involving the application of modern communication technology for the generation of information on varied wild animals that was otherwise not possible to obtain by traditional means opened several new horizons in the field of wildlife research [5]. Radio telemetry solutions for animal tracking using very high frequency (VHF) were initially prominent about four decades ago or so; subsequently, they were succeeded with ultrahigh frequency (UHF)-based satellite telemetry solutions. Later, the global positioning system (GPS)-based tracking solutions emerged and transformed the possible utility of radio telemetry regarding precision and real-time update at a global level. The GPS technology further complemented well with the enhanced cellular communication system [6, 7]. Authors in the [8–10] summarize the advancement in telemetry studies, their use on wide-ranging wild faunal species, and accomplishments in the field of wildlife conservation and at the same time highlight various limitations and reasons for failures.

Based on the radio- and satellite-based telemetry solutions, it has now been possible to bring back some species from the brink of extinction. Additionally, providing much desired research insight and baseline information about them at ease so as to manage their populations effectively, for example, tiger, snow leopard, lion, elephant, rhinoceros, otter, great Indian bustard (GIB), and Amur falcon. In addition to the location/tracking of the animals, telemetry data has helped extensively in carrying out analytics and understanding other significant attributes of the ecology such as habitat–animal interactions, resource use, home range, and migration/dispersal patterns of the animals [6]. Table 8.1 summarizes the number of animals' radio-collared in India across different taxonomic groups during 1983–2011 [9]. Notably, valuable information on seasonal dispersal and migration lacked in the case of GIB (*Ardeotis nigriceps*), a critically endangered bird in India for a long time, in need of telemetry studies.

A recent study [11] on a tagged bustard with satellite transmitter or platform terminal transmitter (PTT), as shown in Figure 8.1, in Maharashtra, India, monitored for 11 months while satellite telemetry tracking for nearly 9 months has revealed that the bird has spent 4 months in a grassland habitat within a PA and 7 months outside the PA (Figure 8.2). Areas frequently used by the tagged bird were actually fallow lands, specifically spared and left unused by villagers for allowing grazing by livestock, and some of the land was fallow due to the absence of irrigation facilities.

Ground tracking revealed that the bird used these areas for resting and foraging and as a cover to avoid disturbance/predation. The crops grown in GIB-frequented areas were primarily rain fed, and the farming practices were less intensive and traditional, indicating it to be bustard friendly. The farmers have unconsciously contributed to bustard conservation through traditional farming

Table 8.1 Summary of radio telemetry studies conducted in India from 1983 to 2013.

Taxonomy group	Studies	Species	Animals tagged	Collars used
Mammals	57	28	314	321
Birds	14	11	55	55
Reptiles	10	7	113	119
Fishes	1	1	1	1
Total	**82**	**47**	**483**	**496**

(a)

(b)

Figure 8.1 (a) Male Great Indian Bustard without PTT tracked in Maharashtra, India. (b) Male Great Indian Bustard with PTT tracked in Maharashtra, India.

practices and keeping land vacant. GIB sightings and satellite telemetry data during 9 months of satellite tracking received 2060 locations. The bird traveled a distance of 1600 km flight to use 7774 km^2 of the landscape; five intensive use sites identified have indicated that in addition to PAs bustards use a large area of human-dominated landscape outside the PAs. It is, therefore, imperative to protect and conserve habitats within and outside the PA. Small land parcels as potential habitat need to be protected across the whole landscape that consists of breeding and nonbreeding sites rather than sole protection of only one

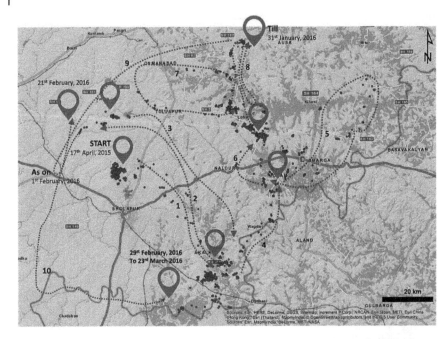

Figure 8.2 Satellite tracking of male great Indian bustard in Maharashtra, India, and its dispersal in human-dominated landscape over a period of 9 months (locations on different dates and routes adopted are shown in sequential numbers).

breeding site. Hence, to effectively conserve GIB population in Maharashtra, a landscape-level approach with multipronged strategy is required.

Satellite imagery of PAs and other natural habitats (referred as remote sensing) globally has transformed the understanding of vital habitats by mapping them, predicting the likely changes to take place in the course of time. In the initial days, the remote sensing data was available for limited locations that are at too low resolutions and a much longer interval; today most natural habitats and PAs in the world are mapped periodically at multiple higher resolution levels. Besides the capture of imagery from satellites (remote-sensed data) can be clubbed together with other spatial features of a geographical location covered broadly within the technological tool referred as geographic information systems (GIS) [12]. Using GIS, it is now possible to study changes in habitat use, the progress of conservation activities, and population distribution of species of concern by superimposing relevant information in "layers" to create intelligent knowledge databases about a given geographical area [12]. Monitoring the location of the animal in context with the geographic location and landscape has led to the development of the concept, that is, *geo-fencing* or *virtual fencing*. Based on the concept, it is possible to avert a potential danger to the animal possibly by venturing in areas that could be hazardous to itself or to humans, for example, animal venturing outside the PA into human settlements [13]. Similarly, no changes in the location of the animal or non-reporting of an expected event at a given point of time could possibly mean that the immobility of the animal is either due to health issues/injury or possible death.

Similarly, capturing animal images of nocturnal, shy, and evasive species using camera traps has been a promising approach for conservation as it helps in monitoring the well-being of an animal through physical recording compared with movement/location tracking. The approach also assists in knowing animal-specific details and thereby carries out closer study on the attributes of a particular individual [12]. Camera trap-based images also help in doing a much effective population estimation of a particular species. Similar to camera traps, acoustic traps have also emerged capable of recording the occurrence of a particular sound, for example, gunshots. Similarly, there is a major research thrust on identification and recording the calls of the animals (especially birds) and a possibility to decipher the communicated message and other behavioral traits [12].

Sensor-based monitoring systems have emerged to monitor diverse physiological parameters of animals ranging from body temperature to heart rate and stress. The sensor systems for monitoring animals are usually referred as sensor tags or electronic tags [14]. The tags of both types, invasive and evasive, have been stated and validated with field-based trials. They can monitor the animals with respect to diverse physical parameters and note their response to ecological changes. The influence of human activities on animals can also be recorded, for example, influence of ecotourism on breeding sites of yellow-eyed penguins [15].

Based on the enormous potential and evident immense scope of biotechnology use in the wide range of fields of wildlife conservation, a large number of DNA sample libraries have come up, especially preserving the genetic code of the most endangered species on the planet. Databases of collected samples make it feasible to genetically barcode a species and identify a given sample [12]. This was almost an infeasible task especially in monetary terms to carry out a couple of years back. The genetic barcoding is helpful and proved handy in determining the details of confiscated contraband material from smugglers and poachers, for example, identifying wild plant and animal species, particularly antelope species used in traditional Chinese medicine [16].

8.3 Challenges: Technology-Based Wildlife Conservation

The gradual technological progression as described in the preceding section has come a long way. However, the available technology and their potential applications in wildlife research, monitoring, and conservation arena have been flawed by various limitations and constraints. Authors in Ref. [8] highlight that the rapid pace of the development of satellite wildlife tracking tools has left little time for thorough testing of new equipment and identifying possible sources of technical failures. They employed 98 satellite collars on 45 Asiatic wild asses (*Equus hemionus*), 334 Mongolian gazelles (*Procapra gutturosa*), 15 Przewalski's horses (*Equus ferus przewalski*), eight wild Bactrian camels (*Camelus ferus*), and two wolves (*Canis lupus*). Although, authors collected valuable data from little-known species in a remote environment of Gobi and Eastern Steppe region of Mongolia, of 98 collars deployed, only 29 worked while the remaining 69 were subject to a wide range of technical problems (human error during manufacturing or deployment, software bugs, mechanical failures, poor GPS performance, premature failures, etc.). Similar experiences of considerable failures for varied reasons have been reported by other fellow workers. Radio implants in the body of small aquatic wild otters require a cumbersome process of chemical capture and risky operation/surgery of endangered species. Nevertheless, over a period, several issues related to large-size transmitters and heavy collars, battery life, and convenient detachment of collars from animal body on completion of studies have been adequately addressed, and undoubtedly large-scale improvements in radio/satellite collars have been made.

The challenges for wildlife conservation are just growing more intense day by day. The foremost reason for growing complexity is the increasing human pressure and the induced interferences in natural ecosystems. The PAs are under continuous pressure and varied threats so as to fulfill human requirements for timber, transportation, mining, power generation, and others. The authors in Ref. [17] discuss the impact of wind turbine areas on migratory birds and the birdlife in the area. Similarly, large structures such as telecom towers and transmission lines induce predator risk and sense of fear in birds, leading to possible habitat loss [18]. Authors in Ref. [18] further state that influence of tall structures is still not well established and causal mechanisms are needed for effective project siting and conservation planning to take place. The impact of ecotourism on breeding of birds has been stated earlier about through utilization of electronic tags.

In most developing countries, the PAs are usually scattered within human-dominated landscapes. Human dwellers within PAs usually referred as forest dwellers or tribal people depend on natural resources for their sustenance. Way back in time, they were considered as a major cause of concern by the authorities as any human activity within the PA is not desirable, and a suggested strategy was to relocate and rehabilitate native and indigenous people [19]. The visionary wildlife conservationists converted this potential problem into the sustainable livelihood (SL)-based approach to conservation by linking local people with conservation efforts and other stakeholders [20]. Since the local people are the most familiar with the terrain and the forest way of life, they can prove to be most useful serving as watchful eyes for any illegal activities and intrusions in the PA.

Animals tend to move out of the legally and administratively defined PAs for a variety of reasons such as extending their home range to their traditional breeding grounds, prey scarcity, and the overabundance of the species within the PA and resultant spillover or search for new territories. This is a typical behavior in the case of large predators. This justifies the significance of the assessing the overall landscape of the region, and the enormous importance to landscape-based approach to conservation is accorded. The changing landscape results in a mosaic situation with a juxtaposition of natural habitats/PAs with human-dominated landscapes. The landscape management approach to wildlife conservation is necessitated in such settings [20, 21]. Authors in Ref. [22] present a landscape-level assessment considering social, physical, and biological attributes of a PA or natural habitats that are also inhabited by transhumant pastoralists rearing livestock. They recommend that grazing activities could be practiced by inhabitants, subject to the pastoralists following the principles of spatiotemporal use of land for livestock grazing. This study justified that mutual coexistence of wildlife and forest inhabitants is feasible. Increase in edge area of the PA due to fragmentation of landscape can result in loss of habitat and connecting corridors for animal movement. These in turn result in one of the most challenging and complex problems of wildlife conservation today, that is, the undesired interaction of wild animals and humans (referred as human–wildlife conflict). However, this term itself has been considered to be ambiguous and subject to misconstrued context of the actual problem [23, 24]. There is a critical observation to make here, that is, human–wildlife interaction gets undesirably close due to human-induced activities and the mosaic landscape situation though the technological systems can monitor the animals using telemetry and electronic tags while they have no bearing on the human-induced changes in the landscape. Similarly, remote sensing and GIS solutions can monitor and report only the resultant changes in the landscape. Further, there could be changes in boundaries/demarcation points that could be reported by these systems; whether they are further pursued for possible encroachment and wrongdoing depends on the personnel responsible for protection of the wildlife areas or conservation reserves. This applies in the case of virtual fencing/geo-fencing.

Apart from the undesired interactions of human and wildlife that can cause serious damage and casualties, there are other aspects of the interaction, such as that on seeing an animal with a device attached could lead to displeasure among people. The authors in Ref. [25] present a complete study on the interaction between stakeholders and electronic-tagged aquatic animals. Seeing an animal injured due to a mounted device could lead people to indulge in helping the animal, which could possibly cause further harm to the animal as they are untrained to handle such a situation. Therefore, technological solutions for the wildlife monitoring are required to take into account the sensitivities of all the stakeholders.

This brings up another important observation about wildlife conservation using technology, that is, majority of the mechanisms involve complex equipment and procedures. Often, the frontline wildlife conservation staff are capable of handling only simple technical equipment and procedures. The ICT-based wildlife conservation solutions often require a high-level expertise to manage the equipment/devices and derive productive use from them. The technology-based solutions operate only under certain environmental and operational conditions, for example, the radio collar-based tracking requires the animal to move in less scattered and not highly dense areas; else the radio communication is likely to be lost. Similarly, the event of a sudden shock to the device due to an abrupt physical activity by the animal could render the collar ineffective for use. The possible repair or replacement of the device could necessitate a tedious process of tranquilizing the animal. Therefore, at the moment, the reliability of the technological solutions for wildlife conservation has serious limitations.

The majority of the technological solutions for wildlife conservation are expensive and commonly involve costly procedures for implementation and maintenance. Interestingly, majority of rich habitats and biodiversity hotspots in the world are found in countries that are economically weak (developing or under developed). The exorbitant cost of equipment puts a limitation on the possible use, and it imposes a resource crunch for the wildlife management authorities since the funds have to be diverted from other defined utilities. Possibilities of theft/damage/handling and maintenance of the equipment also impose an economic burden that is often unbearable by the concerned authorities. There is a possibility of misuse of technology for inappropriate tasks, for example, unmanned aerial vehicles or drones meant for surveillance and fitted with sensors for wildlife monitoring could be used for spying on the forest dwellers or monitoring sensitive national establishments [12].

Globally, wildlife conservation efforts are marred by one major human-imposed hurdle that is legal enforcement and legal sanctity of the conservation policies. Often the wrongdoers manage to get away with petty fines and/or minimal jail imprisonment. Additionally, the resources allocated for wildlife conservation agencies are nominal in comparison to resources provided for infrastructure, social welfare, education, and others. The authors in Ref. [26] detail the resultants of noncompliance with conservation issues and discuss the diverse drivers for noncompliance. Likewise, authors in Ref. [24] state that evidence-based decision-making, educating the stakeholders, and proactive enforcement can help bring about control in human–wildlife conflict. Overall, it would be fair to say that the problem is growing complex day by day and it is immensely important to ensure the gap between wildlife conservation policy formulation/implementation, and unlawful activities using modern tactics and techniques by wildlife poachers and smugglers should be minimized. It would be appropriate to mention that the wrongdoers can only be stopped by the effective empowerment of the wildlife conservation staff alone, and probably limitless amount of stand-alone technological solutions to monitor wildlife solely cannot substitute or counter that.

8.4 Possibilities in Future: Technology-Based Wildlife Conservation

Multiple possibilities about the likely condition of wilderness and wildlife could be visualized by the year 2050. The foremost possibility is that most wildlife species might be wiped out and extent of wilderness (terrestrial and aquatic ecosystems) drastically reduced, and they are left as just remnants or islands or sink habitats amid human-dominated and highly modified/developed landscapes. The second possibility forecasted could be that much of the human population has migrated from countryside and rural areas to the centers of urban agglomeration and natural ecosystems once again get an opportunity to regenerate and recover resulting into abundance of at least most common wildlife species except specialist, and the rare, endangered, and threatened (RET) species those have already locally extinct long ago in the process. Most urban wildlife seen in present time must have also declined. Most grasslands have already disappeared in developing/tropical countries. Wetlands including rivers and their floodplains are most threatened entities. There are remote possibilities that humans might be able to develop expertise and require technology to restore and reclaim much of the degraded natural ecosystems. However, it is sure in all probability that the present generation of field/frontline forest and wildlife staff is physically fit and capable of performing arduous professional protection and conservation duties in harsh and difficult terrain. While the new generation may not be interested in difficult outdoor profession, it becomes much more relevant to have appropriate technology applications available for various types of wildlife researches, monitoring, and conservation requirements. Reintroduction and rehabilitation of wildlife will become major activities maybe in the next decade itself. Such conservation efforts will definitely need a variety of technologies and other tools.

Considering the previous scenario, in the future it would be much desirable to track most activities of wild animals, and it would be possible to keep a very close eye on them and their habitats by way of technology advancement based on five human senses and their appropriate integration.

The sensors need to realize a physical entity (subject) through five human senses, namely, sight (seeing), auditory (hearing), olfactory (smelling), gustatory (tasting), and tactile (touching), sensed by five distinct entities, that is, eye, ear, nose, tongue, and skin. Relying on the five senses is a concept of human bond communications [27] that is different from the conventional human-centric communications relying solely on seeing and auditory-sensed information. Advances in sensing technologies will enable monitoring a wide range of animals with high precision (*seeing*). It would be possible to monitor on a regular basis the diet of an animal (*gustatory*). Sensors collectively forming the rich Internet of things in the wildlife habitat/PAs would make it feasible to obtain varied valuable and vital data. Applying analytics on Internet of things data would present a new perspective for the conservationists to visualize and analyze the natural ecosystems and the wildlife. The sensors could be placed on prominent trees, recognizable landmarks within the protected habitat. Animals of a given species have a liking for a particular food item, roaming and staying in proximity to certain given species of trees/shrubs (*olfactory and tactile*), and similarly drinking water/wallowing at only a particular water hole in the protected area. Currently, it is the field personnel who observes and records some of these attributes, but with Internet of things, it would be possible to monitor these attributes in an automated manner. Animals give distress signals and alarms to alert others within their community of an imminent danger; this also helps animals that share a symbiotic relationship (e.g., deer and monkey). Based on the alarm calls also with inputs from other sensory data (radio collars), it would be possible to judge remotely whether the distress call has anything to do with undesired human presence and making it possible to take swift action on poachers, if any. Human presence can also be detected by the sensors inconspicuously placed in the PA wherein the human being is seen (*seeing*) or gets in contact with the sensor (*tactile*). The telecommunication equipment would be capable of picking the olfactory (*smell*) of animal kills and touch (tactile) details of an individual.

Lately, it has been recognized that natural ecosystems are providers of enormous ecological and economic services considering the varied goods and services that are provided by them directly or indirectly from various environmental produce, for example, oxygen, timber, rubber, and carbon sequestration. In future, it would be possible to compute economics for a given natural ecosystem and wildlife reserves on a continuous basis using appropriate technologies, thereby ensuring their better protection by eliciting much desired people support.

In future, it would be feasible to determine the authenticity of wildlife produce in comparison with counterfeit products swiftly utilizing smell/taste (*olfactory and gustatory*) determining sensor-based systems. Currently, the process is complex and expensive for implementation on a large scale to cover all natural areas. Nanotechnology relying on nanosensors would serve this purpose, possibly using portable handheld devices, wherein the nanosensors interact with the particles of the product and determine whether product is counterfeit/imitation or genuine.

8.5 Role of Conservationists in Wildlife Conservation

Based on the foregoing description of the technology advances made so far and their appropriate integration into wildlife conservation efforts, challenges imposed in use of such technologies and future prospects, the significance of wildlife conservationists and field practitioners, in particular, are of utmost importance. Their effective empowerment and utility in serving the purpose is vital for successful conservation. It cannot be overemphasized here that human beings can be the best defense to other human beings, and technological solutions that empower the rightful action by them and prevent wrongdoing by them would go way ahead compared with technological solutions to monitor wildlife. Enriched communications relying on all five senses in the form of human bond will be of immense utility in this regard. The policy formulation and law enforcement should place the wildlife personnel at the forefront, and they should be the sole authority on deciding the best course of action concerning the natural ecosystems, PAs, and wildlife resources. Additionally, the observations based on the data from technological solutions if administered by field personnel should have full legal sanctity and precedence. Another major factor that necessitates involving the conservationists in the effective policy and decision-making and conservation efforts through appropriate technological applications is the need for sharing common mental models [28]. An individual holds his opinion and perspective of a given situation based on his/her area of expertise and profession, for example, the researcher developing an animal tracking system has a different opinion and understanding about the wild animal and its habitat compared with a frontline field practitioner.

8.6 Conclusions

Undoubtedly, there is mounting pressure on the natural ecosystems, wildlife habitats, and wildlife. Technology innovations and appropriate utility have come a major way in addressing the multiple, complex issues relevant to wildlife conservation. However, the technology use is marred with certain critical and complex constraints. In the future, the utility of technology-based interventions for wildlife conservation is expected to grow exponentially, and utilizing principles of human bond communication would significantly benefit it. In order to ensure an optimal technological utilization for wildlife conservation, that is, conservationists and frontline staff should be adequately empowered, clearly demonstrated the potential of technology use and integration so as to ensure the desired outputs. This way they can serve better in order to conserve the natural habitat but also prevent species from extinction. Thus, there is a felt need that the technology advancement, its integration, and appropriate solutions are designed bearing in mind the field conservationists and policy instruments that empower them to take appropriate decisions for conservation efforts.

References

1 Aichi Biodiversity Targets (2010) Convention on Biological Diversity. https://www.cbd.int/sp/targets/. Accessed on Jul 3, 2016.
2 Biodiversity Hotspots (1990) Conservation International. http://www. conservation.org/How/Pages/Hotspots.aspx. Accessed on Jul 3, 2016.
3 S.H.M. Butchart, M. Clarke, R.J. Smith, et al., 'Shortfalls and solutions for meeting national and global conservation area targets', *Conservation Letter.* **8**:329–337 (2015).
4 I. Breckheimer, N.M. Haddad, W.F. Morris, et al., 'Defining and evaluating the umbrella species concept for conserving and restoring landscape connectivity', *Conservation Biology.* **28**(6):1584–1593 (2014).
5 K. Sivakumar, B. Habib (Eds.). *Telemetry in Wildlife Science. ENVIS bulletin, Wildlife & Protected Areas.* Vol. **13**, No.1, Wildlife Institute of India, Dehradun, pp. 246 (2010).
6 A. Balmori, 'Radiotelemetry and wildlife: highlighting a gap in the knowledge on radiofrequency radiation effects', *Science of the Total Environment.* **543**(Pt A):662–669 (2016).
7 J. Wall, G. Wittemyer, B. Klinkenberg, et al., 'Novel opportunities for wildlife conservation and research with real-time monitoring', *Ecological Applications.* **24**:593–601 (2014).
8 P. Kaczensky, T.Y. Ito, C. Walzer, 'Satellite telemetry of large mammals in Mongolia: what expectations should we have for collar function?', *Wildlife Biology in Practice.* **6**(2):108–126 (2008).
9 B. Habib, S. Shrotriya, K. Sivakumar, et al., 'Radio-telemetry studies in India—issues and way forward', In Sivakumar K and Habib B (eds) *Telemetry in Wildlife Science. ENVIS Bulletin, Wildlife & Protected Areas.* Vol. **13**, No.1, Wildlife Institute of India, Dehradun, pp. 3–19 (2010).
10 S.A. Hussain, 'Application of radio-telemetry techniques for studying ecology of otters and related small carnivores', In Sivakumar K and Habib B (eds) *Telemetry in Wildlife Science. ENVIS Bulletin, Wildlife & Protected Areas.* Vol. **13**, No. 1, Wildlife Institute of India, Dehradun, pp. 111–116 (2010).
11 B. Habib, G. Talukdar, R.S. Kumar, et al., 'Tracking the Great Indian Bustard in Maharashtra, India', Technical Report 2016. Wildlife Institute of India and Maharashtra Forest Department, Dehradun and Nagpur. pp. 20 (2016).

12 S.L. Pimm, S. Alibhai, R. Bergl, et al., 'Emerging technologies to conserve biodiversity', *Trends in Ecology & Evolution.* **30**:685–696 (2015).

13 D.S. Jachowski, R. Slotow, J.J. Millspaugh, 'Good virtual fences make good neighbors: opportunities for conservation', *Animal Conservation.* **17**:187–196 (2014).

14 A.D.M. Wilson, M. Wikelski, R.P. Wilson, et al., 'Utility of biological sensor tags in animal conservation', *Conservation Biology.* **29**, 1065–1075 (2015).

15 U. Ellenberg, T. Mattern, P.J. Seddon, 'Heart rate responses provide an objective evaluation of human disturbance stimuli in breeding birds', *Conservation Physiology.* **1**(1):cot013 (2013).

16 J. Chen, Z. Jiang, C. Li, et al., 'Identification of ungulates used in a traditional Chinese medicine with DNA barcoding technology', *Ecology and Evolution.* **5**(9):1818–1825 (2015).

17 W.P. Kuvlesky, L.A. Brennan, M.I. Morrison, et al., 'Wind energy development and wildlife conservation: challenges and opportunities', *The Journal of Wildlife Management.* **7**1:2487–2498 (2007).

18 K. Walters, K. Kosciuch, J. Jones, 'Can the effect of tall structures on birds be isolated from other aspects of development?', *Wildlife Society Bulletin.* **38**:250–256 (2014).

19 V. Reid, 'Wildlife and human-friendly conservation: a contradiction or a possibility?', *Biodiversity.* **16**:1–2 (2015).

20 P.K. Mathur, P.R. Sinha, 'Looking beyond protected area networks: a paradigm shift in approach for biodiversity conservation', *International Forestry Review.* **10**(2):305–314 (2008).

21 N. Midha, P.K. Mathur, 'Assessment of forest fragmentation in the conservation priority Dudhwa landscape, India using FRAGSTATS computed class level metrics', *Journal of Indian Society of Remote Sensing.* **38**(3):487–500 (2010).

22 B.S. Mehra, P.K. Mathur, 'Livestock grazing in the Great Himalayan National Park conservation area—a landscape level assessment', *Himalaya, the Journal of the Association for Nepal and Himalayan Studies.* **21**(2):89–96, Article 14 (2001).

23 S.M. Redpath, S. Bhatia, J. Young, 'Tilting at wildlife: reconsidering human-wildlife conflict', *Oryx.* **49**(2):222–225 (2015).

24 S. Baruch-Mordo, S.W. Breck, K.R. Wilson, et al., 'The carrot or the stick? Evaluation of education and enforcement as management tools for human-wildlife conflicts', *PLoS ONE.* **6**(1):e15681 (2011).

25 N. Hammerschlag, S.J. Cooke, A.J. Gallagher, et al., 'Considering the fate of electronic tags: interactions with stakeholders and user responsibility when encountering tagged aquatic animals', *Methods in Ecology and Evolution.* **5**:1147–1153 (2014).

26 J.N. Solomon, M.C. Gavin, M.L. Gore, 'Detecting and understanding non-compliance with conservation rules', *Biological Conservation.* **189**:1–4 (2015).

27 R. Prasad, 'Human bond communications', *Wireless Personal Communications.* **87**(3):619–627 (2016).

28 D. Biggs, N. Abel, A.T. Knight, et al., 'The implementation crisis in conservation planning: could "mental models" help?', *Conservation Letters.* **4**:169–183 (2011).

9

An Investigation of Security and Privacy for Human Bond Communications

Geir M. Køien

Faculty of Engineering and Science, Department of ICT, University of Agder, Kristiansand, Norway

9.1 Introduction

This chapter provides an initial analysis and investigation of security and privacy issues for human bond communications (HBC). Immediately, we may ask: what do we mean by HBC? As a first attempt, we may define HBC to be communications that capture and communicate what we may describe as the input to the human sensory apparatus. Later, we will broaden our perspective considerably.

The senses Traditionally, we have the five senses: *sight, hearing, taste, smell,* and *touch.* We may add that humans have awareness of balance and motion; we can also feel pain, temperature, and pressure.

A prerequisite for HBC is that the physical experience that influences a sense can be captured digitally. For completeness' sake, one also ought to be able to capture input from all of the human senses. Having assumed that, all we need is a means to communicate the data and convey them to the target person. This is no mean feat, but it is also something that is quite conceivable and well within an extended virtual reality (VR) vision.

Circumventing the senses Sensory input is communicated to the brain via the neurons in the sensory organs. So, it seems that the next logical step to the HBC concept would be to circumvent the senses. A logical extrapolation to this is therefore to let the communications be performed brain to brain. Then one needs to capture and convey the sensory experience or mental state at the brain itself. That necessitates brain extensions, at least with respect to the interfaces. We may then talk about brain-to-brain communications or even mind-to-mind communications.

Human Bond Communication: The Holy Grail of Holistic Communication and Immersive Experience, First Edition. Edited by Sudhir Dixit and Ramjee Prasad.
© 2017 John Wiley & Sons, Inc. Published 2017 by John Wiley & Sons, Inc.

Virtual brain and mind communications Having gone that far, the next level is of course to dispense with the physical human brain altogether. This allows for running the mind on a computational substrate that runs, or emulates, a virtual brain. This brain may be an enhanced and extended brain with much higher capacities. This, of course, will fundamentally influence our thinking of who we are.

Now, our perception of self is subjective and it is even said to be an illusion, a social construct, created by the mind [1]. Of course, the concept of the mind is no less subjective, and the late Marvin Minsky popularized the notion of a "society of mind" [2]. We shall return to these topics later.

The computational substrate may or may not run virtual brains; that is, the computational substrate may run nonhuman minds that do not require a virtual brain to run on. Those nonhuman minds will be artificial intelligences (AIs), but, of course, at that stage, one may discuss if there are any non-artificial minds left anyhow.

9.1.1 The Unknown Future Rolls Toward Us

9.1.1.1 Futuristic Outlook

This paper is a scientifically qualified speculation on what we may expect to see if we extrapolate current technological trends in information and communications technology (ICT). The approach taken is not to investigate concrete technological aspects *per se* but rather to convey a vision of what HBC will likely bring about for us personally and for our society. And that vision includes not only communicating sensory information from person to person but also communicating memories and experiences directly from brain to brain. And, of course, as the paper title indicates, we shall investigate what this may mean in terms of security and privacy.

9.1.1.2 Enlisting Science Fiction and Opinion Essays

The quote "The unknown future rolls towards us" is from the movie *Terminator 2: Judgement Day* [3]. It points to another field with speculation about the future, namely, that of science fiction (SF). At its best, the SF genre has the power to point at dilemmas and issues, sometimes regarding the future and sometimes, often in disguise, pointing to more contemporary issues. For the purpose of this paper, we shall take the liberty of invoking SF visions of the human condition in a future where technology has the potential to radically change us humans and our society, and ultimately even destroy us. SF is thus useful for highlighting what the technologies, such as HBC, may mean to society. SF may therefore provide useful perspectives on HBC-like threats, which is relevant for a paper on HBC security.

We will also cite and refer to opinion essays by well-known professionals, who express their opinions regarding their fields. An example here is Bill Joy's "Why the Future Doesn't Need Us" article [4]. Needless to say, such opinion articles are often criticized, but then again, they were probably written to start a debate.

9.1.1.3   Long-term View

We take a long-term view in the sense that the technologies we presume need not be available today. However, the working hypothesis is that there is no fundamental obstacle to realizing the technologies. That is, we presume that there will be no insurmountable technological limits that prevent the scenarios from unfolding. Thus, it then follows that if there are no fundamental limits, then the developments anticipated in this paper are possible and may very well materialize in one form or another.

The "long-term" aspect of these speculations is not truly long term. Given the tremendous technological progress we have experienced within the field of ICT during the last decades, we must be prepared that the "long-term" view may be experienced by our children (or grandchildren). So, in a historical context, the view is really a short-term view.

9.1.2   Security and Privacy

The concept of directly communicating the input/output of the human senses obviously poses some challenges with regard to security and privacy. The issues will be even more acute for the extended scenarios. We will discuss this further in Section 9.5.

9.1.2.1   Security and Assurance

In practical terms, there is a need for *assurance* of the sensory input. As always in security, we need to establish who is involved (principal entities, intruders, and other parties), determine what the assets in question are, quantify what value the assets have, and find out what the threats may be, what the vulnerabilities are, and so on. In addition the different parties involved may have different opinions regarding the assets, the value of the assets, the threats, etc.

9.1.2.2   The Who Question

In security parlance the "who" question is about who the principal entities are. So, we need to identify the principal entities, noting that an entity may be any autonomous agent. This means that, we already have the language needed to describe nonhuman parties. The principal entities will need to be identified, and for this purpose one may use designated identifiers or even by-association information elements like an address.

9.1.2.3   The Trust and Accountability Question

It is all very well to be able to identify other entities, but somehow we need assurance concerning the honesty and trustworthiness of the entities. Human trust in humans is not infallible in any way, but we have the option of resorting to higher authority (law and order) and thereby make the other entity, the claimed culprit, accountable. In an HBC reality these issues will likely be less clear, and it is not obvious at all how trust will arise and be handled, or how accountability (law and order) will be handled.

9.1.2.4 The Asset Question

It matters a great deal what the assets are. The assets are, after all, the objects (information) that one wants to protect. For HBC, the asset question is easily answered. The information contained in the "HBC experience" is the asset.

In the extreme, one may envision capture of the entire brain state (or mind state). In that situation it is not obvious that there is a difference between the "Who" and the "Asset." So, potentially, the self ("I") is the asset.

9.1.2.5 Privacy Matters

The fact that some piece of data may need protection and secrecy does not by itself imply that the data are privacy sensitive. For HBC data, the privacy aspect is an intrinsic property in that the data is captured from a person. However, even in this case, individual readings may be anonymized. What makes the data privacy sensitive is that the data is intrinsic to a specific individual and/or that it is linkable to a specific individual. Terms such as anonymity and unlinkability have specific meanings within the privacy field. The terminology can be confusing, and we refer to Ref. [5] for a comprehensive overview.

Privacy-sensitive data will invariably need confidentiality protection. Privacy is also about control of personal data. Not only do we need control over who sees the data (read permission), but we must also have control over who can use the data (execute permission) and who can create/amend/modify (write permission) the data. This extends further, and the delegated rights may be subject to various conditions and restrictions. This can include purpose of access and various spatio-temporal conditions.

9.1.2.6 The Right to Your Own Experiences

For HBC one may envision an intellectual property right (IPR) dimension too. That is, if you originate an experience, you may be entitled to legal rights to the experience. We shall completely avoid any legal analysis in this paper, but provided that IPRs are recognized, then one will need methods for IPR tracking and enforcement. It is of course not at all clear to what extent IPRs will make sense in an HBC context.

9.2 Fundamental Assumptions, Premises, and Issues

9.2.1 Brain Versus Mind

We need at times to differentiate between brains and minds, but what exactly is the difference? To cite Marvin Minsky, "the mind is what the brain does" [2]. This is indeed a good starting point.

To this end, the brain is seen as a complex and highly adaptive biological computational substrate. The mind is somewhat akin to the "software" instantiated on the computational substrate (hardware), but given that the brain more or less continually "rewires" itself, the analogy is at best only partially accurate. Principally, we postulate that a mind may be run on any sufficiently capable and flexible computing substrate. A mind running on a non-brain substrate may in principle be an upgraded human mind, but it may also be an AI. The adjective "artificial" is of course relational, and it begs the question on what we really mean by antonyms like "real" and "genuine" in this context. Taken together, it is clear that this points to a Strong AI view, but not all HBC visions depend on a Strong AI assumption.

9.2.2 Strong AI

Philosophically, one may question whether Strong AI is for real, since it requires understanding (intentionality) and consciousness. These concepts are ill-defined and not truly useful for classifying intelligence. The view that Strong AI does not make sense is famously presented and expressed by Searle in his "Chinese room thought experiment" [6], where he refutes Strong AI as such.

However, indistinguishability may be all we need. If an entity is indistinguishable from an intelligent entity, then the argument goes; we may as well consider it to be intelligent. This argument is a logical extension of the Turing test [7], and indeed Turing himself argued that "instead of arguing continually over this point it is usual to have the polite convention that everyone thinks" [7]. Turing himself obviously extended his "polite convention" to sufficiently capable machines. Turing also, quite obviously, believed that thinking machines was possible [8]. I.J. Good, a wartime colleague of Turing, extended the notion and promoted the idea of an "intelligence explosion" in conjunction with thinking machines [9]. Good also held the view that once a machine reached a certain intelligence level, it would almost immediately surpass human intelligence and become far superior to us.

With respect to cognitive capacity, it is obvious that the human cognitive circuitry is very slow and there are definite upper limits to the number of nerve cells in our neural system and brain. Complete emulation of our cognitive capabilities at the cell level is therefore in principle possible. Computationally it may or may not make sense to directly emulate brain functionality, but there seems to be no fundamental obstacle to do so in the near future. Futurologist Ray Kurzweil [10] is a principal proponent of this view. Other scientists, for instance, Gary Marcus, an esteemed author and professor of psychology and neural science, have come somewhat hesitantly to the conclusion that the brain is a computer [11]. According to Marcus, the question is what kind of computer the brain is (from *The Computational Brain* in Ref. [12]).

Since the term Strong AI is emotionally charged, one may take the position that indistinguishability is all that is needed.

That is, if one can emulate a human mind to such an extent that one cannot determine if the mind in question is running on an artificial substrate or not, then we might as well not care.

9.2.3 Computability, Meaning, and Intelligence

One view in modern physics is that everything is about computability. In this world view, the universe is a computer that creates/computes its own reality. The fundamental building blocks, whether they be quarks, strings, or whatever, are essentially atomic computing elements. Here we note that any physical "operation" that can be construed as a logical operation are indeed performing computations. So, computation may ultimately be *the* intrinsic characteristic of a physical reality.

So, then, what does the universe compute? Apparently, the answer is "itself." This view is complemented by the "Created Computed Universe" school of thought [13]. By extension, if the universe is created as a computational model, then we may be simulations. Bostrom has proposed the simulation argument [14] that basically expresses the view that we live in a computer simulation.

> This paper argues that at least one of the following propositions is true: (1) the human species is very likely to go extinct before reaching a post-human stage; (2) any posthuman civilization is extremely unlikely to run a significant number of simulations of their evolutionary history (or variations thereof); (3) we are almost certainly living in a computer simulation. It follows that the belief that there is a significant chance that we will one day become posthumans who run ancestor simulations is false, unless we are currently living in a simulation. A number of other consequences of this result are also discussed.
> —from the abstract of "Are you living in a computer simulation?" [14]

The simulation point of view does not obviate the need for a bottom layer in the simulation, and so the base case is still unresolved in this world view.

9.2.3.1 Meaning
"Meaning" is obviously an abstract philosophical concept. It cannot be directly associated with the atomic computing elements but is rather an emergent concept. That is, meaning is an emergent acquired characteristic. It may even be that meaning is unavoidable for a sufficient complex dynamic system, thus making it an intrinsic property of such systems.

9.2.3.2 The Inevitability of Intelligence

Does "meaning" have any meaning without an intelligence to ponder over the meaning? We mean this to be a rhetorical question, since we don't think it can be answered without invoking self-reference. Still, one may speculate that when a dynamic system reaches a threshold, spontaneous order and pattern replication may be unavoidable. From then on, emergence of intelligence may be a given if the system is allowed to grow and exists for a sufficiently long period. We may very well be proof of that.

However, in this chapter it is of little consequence to us here if intelligence is inevitable or not. All we care about is whether an already intelligent entity (unproven proposition) can create, or at least instantiate, another intelligent entity. The instantiation part is related to uploading of minds to virtual brains or some suitable computational substrate.

9.2.4 Levels of HBC

There are several possible levels of HBC. For the purpose of this paper, we distinguish between two "ordinary" principal levels and one extraordinary level. These are the following:

- **Weak HBC**: Here we have that sensory input from the human senses is captured at person A, transferred to person B, and then instantiated as sensory input to the senses of person B. This version is quite close to what may be achieved by a complete version of current VR technology.
- **Full HBC**: In this version, we bypass the whole of the sensory apparatus and capture and instantiate information directly in the brain. That is, we capture, transfer, and instantiate brain state. Here we positively require brain interface augmentation. Note that the brain state *may* be artificial or augmented.
- **Deep HBC**: In this version, we bypass the requirement for a human to be involved. That is, we abandon the requirement for a biological brain. There might be a nonbiological brain or a virtual brain, both capable of running human minds. Then of course, the mind in question may also be a completely artificial mind. When nonhuman minds are involved, one may call this *mind communications*.

Even for **Weak HBC**, there are potentially enormous consequences for the persons involved and for society at large. Many aspects of our culture would be totally transformed, and over time the need for travel and so on would be greatly diminished. Of course, the potential for abuse is ever present.

Full HBC has the potential to eradicate the "I" and fundamentally change how we, as a species, interact. For one, learning will be as easy as uploading the required knowledge. Similarly, Full HBC will enable false memories a la those in the movie *Total Recall* [15], which was based on a short story by Philip K. Dick [16]. Needless to say, but this clearly has the power to change us and our society in ways we can only begin to comprehend.

The **Deep HBC** level more or less presupposes a Strong AI view, although one may remain agnostic and ignore the issue of AI as such. Deep HBC will involve our minds being uploaded to nonhuman computational substrates, which by itself will radically change the human condition. Our new brains would potentially be vastly more powerful, if for no other reason for the fact that the new computational substrate may run millions of times faster than our biological circuitry. Deep HBC also easily invites hugely enhanced capacity in terms of increasing the number of computing units (artificial neurons or whatever atomic unit there is). Taken together, a "human" mind running on a nonhuman computational substrate would be a very different entity from that of a human of today. Certainly, Deep HBC also invites AI entities, but it is not clear how meaningful the distinction between Deep HBC humans and AI entities will be.

The ethical dilemmas will run very deep and it is not at all clear how these will be resolved. The power of a fully developed HBC technology will seriously challenge our concept of personal integrity and ultimately also challenge the very notion of "I."

However, in this paper, we take a somewhat "simplistic" view of what one can/must do in terms of technologically based security. That is, given that HBC will highlight personal privacy and security issues, what are the technical challenges that must be addressed to meet the issues? We will not attempt any concrete solutions in this paper; we will merely strive to get the questions right.

9.2.5 Created Computed Universes

Of course, if one can envision copying brain states and running them, what then prevents a fully simulated/emulated universe?

If one takes into account the universality of computation and views physics as computations, then what prevents a universe(al) Turing machine from *being* the universe? The "Created Computed Universe" ideas are not new. An account is found in the article "Created Computed Universe" [13]. This may appear a little like the reality in *The Matrix* movie [17], but as Bostrom noted in Ref. [14], we may actually live in a simulation.

9.2.6 Exponential Progress

The advent of the Weak HBC vision is decidedly a near term prospect. The Full HBC vision does require much more, and the Deep HBC vision is for a future where we truly ascend our current humanity. So, the question is can it happen at all? And, if so, when will it begin to happen?

Technological progress seems to be accelerating. We have Moore's law and so on, which seem to indicate that computing capacity is exponentially increasing. There are other similar "laws" that also point to exponential progress. The general argument behind these observations is that the progress made, say in computing capacity, is also applied to the process of improving computing capacity. What we have is that we continually make better tools, and these tools are themselves used to make even better and more powerful tools. Kurzweil has even proposed this as a "law of accelerating returns" [18]. Exponential growth cannot continue forever with a finite pool of resources. So, progress in increasing our computational capacity will eventually flatten. Or, we will have to continually utilize more and more resources.

With respect to reaching the threshold of being sufficiently capable for simulating/emulating a human brain, there is no shortage of large-scale projects that directly work toward this goal (the Human Brain Project (HBP) [19] is but one prominent project in this area). It is therefore fairly uncontroversial to assume this to happen within the next few decades.

Progress in areas such as quantum computing may make this happen much faster. However, in a historical context, whether it happens within a few decades or a few centuries matters little; the only thing that matters is whether it can be done or not. And there appears to be no obvious fundamental obstacle to prevent a Deep HBC vision to emerge. If so, it seems a safe bet that it will be realized, whether it be in this century or later.

9.3 Human Bond Communications

9.3.1 Our Senses and How to Capture and Reproduce the Data

Let us, for simplicity, limit ourselves to the five senses: *sight, hearing, taste, smell*, and *touch*. How do we capture what our sense captures?

9.3.1.1 Sight

The camera technology to capture what our eyes can capture is already here. There are practical issues with equipment size, with focusing, and others, but essentially, our technology is already performing *better* than our eyes in a number of situations. We can expect this trend to continue, and we should expect technological sight capture to seriously outperform our eyes in the future. This may allow "seeing" in infrared or ultraviolet, if we so desire. The bandwidth requirements to merely reproduce what our eyes can capture are in the order of 10 Mbps [20]; enhanced vision will require higher bandwidths.

Reproduction of the data can range from simply viewing a reconstructed picture or movie to bypassing the eye and feeding the input directly to the visual cortex. There is substantial filtering and processing done by the visual cortex, and this must be accounted for in the way the interactions are carried out.

9.3.1.2 Hearing

To capture hearing seems simple enough. We need sufficiently good microphones, but capturing the 20–20,000 Hz range is not a challenge, and the bandwidth requirements are modest. Good audio codecs exist, and better ones will be developed. That is, even for very high fidelity, we should expect them to require no more than a couple of megabits per second bandwidth.

Reproduction can be done by a loudspeaker and using our ears. We may also bypass the ears and feed the input to the auditory cortex directly. We already have cochlear implants (CI), where the CI bypass the normal hearing process and stimulate the cochlear nerve directly. An account of CI and future developments is found in [21]. Similar to the visual cortex processing, the auditory brain system performs both basic and higher-level processing functions, and this must be accounted for in the reproduction stage.

9.3.1.3 Taste

Taste can be categorized into five basic tastes: sweetness, sourness, saltiness, bitterness, and umami. To differentiate between these tastes, we have in the order of 5,000–10,000 taste buds. Taste is not instantaneous and the data throughput is relatively low. The challenge for capturing taste is therefore more on the capturing part than data transmission. Technologically based taste reproduction is as yet fairly unexplored territory, but there is no reason to assume it cannot be done, and it will obviously be possible to duplicate the neural response of the taste buds.

9.3.1.4 Smell

Smell is related to taste, but the olfactory receptors are clearly independent of the taste buds. That is, the sensors need to be targeted to smell, and they need to be sufficiently sensitive. Some animals have a much better sense of smell than we do, and these animals probably have better smell receptors, more of them, and a larger set of neurons to convey the smell to the brain. The bandwidth requirements are modest. Technologically based smell reproduction is as yet fairly unexplored territory, but there is no reason to assume it cannot be done, and it will obviously be possible to duplicate the neural response of the olfactory receptors.

9.3.1.5 Touch

Touch is complex and challenging to cover. To develop sensors for touch essentially means capturing input from the whole of our body surface. Our skin has receptors for heat, cold, touch, and pain. For an average size human, this translates to something in the order of 2.7 million nerve endings, with the potential to fire about 50 times per second. In total we approach a peak capacity of about 135 Mbps, but in truth it is unlikely that our brain is really expecting this type of load. Still, a comprehensive touch sensory apparatus may need to capture in the order of a couple of hundreds of megabits per second. Technologically based touch reproduction is in principle not difficult at all, but to reproduce it for the whole of our body does pose some challenges. It may be easier to intercept nerves and induce the required signal there.

9.3.1.6 Complete Capture

To capture all that our sensory apparatus captures will require developments and technological progress, but there does not seem to be any fundamental obstacle to any of this. The communications capacity required will be in the order of a few hundred megabits per second, which is feasible even today.

9.3.2 Sensory Data and Information Processing

The human sensory inputs are heavily processed by the brain and our nervous system. So, the human user experience is dependent not only on the actual sensory inputs but also on our perception of the inputs. Some of the perception filters will be more or less hardwired into our nervous system and brain, like our basic face recognition ability, while other filters are learned or enhanced through learning. Our expectations also color the processing significantly, and it is fairly obvious that the processed data is something quite different from the raw data input from our senses.

So, what does it then mean to communicate the sensory inputs? Obviously, for sight and hearing, we have no problem accepting pictures and sound. While seeing a picture is not quite the same as seeing the sight, our brains are quite flexible and allow us to get a substantial part of the experience of a sight through seeing a picture. Sufficient resolution plays a part here. For low resolution pictures, we will be more aware that it is a picture, but even so people have been watching low resolution television for years and listening to phone conversations where the actual quality is far from the real thing. So, from a technical perspective, the question will perhaps be what the *necessary* resolution is.

Raw sensory data The previously mentioned examples demonstrate communicating the "raw" inputs. The perception processing is not contained in the picture or the captured sound. Art may be different, and with artistic expressions, one may see that as a processed object. Still, perception processing is nevertheless done in the brain of the person viewing the object or listening to the music.

Processed sensory data We may of course take the more radical view to HBC, where we communicate the *perceived* image or sound experience. That is, we communicate directly the experience felt by a person. This could encompass inputs from all sensory organs and whatever perception processing that the original perceiver did. That is, one uniquely and completely shares an experience or, more likely, an approximation of the experience.

9.3.3 The Capture–Communicate–Instantiate Challenge

To convey experiences, we need to *capture, communicate,* and *instantiate* (CCI) the experience information. A CCI sequence is in principle independent on whether it applies to the raw sensory input/output, the perceived/processed data (brain state), or even the mind state.

Irrespective of HBC level, the CCI challenge can be summed up as:

- **Capture** the data from an entity (read/copy).
- **Communicate** the captured data (transfer).
- **Instantiate** the data in another entity (write/run).

Capture and instantiate It is not at all clear how to capture an experience and encode it. To capture the raw sensory data is in principle less complicated, and the amount of data is contained by the communications capacity of the neurons. It is also not clear how to instantiate an experience in another brain/mind. In contrast, to instantiate input from the sensory system, while a complex and technologically highly challenging operation, does seem readily feasible.

Communicating The communications part itself is comparatively trivial. For communicating sensory input, we are already capable of communicating at these speeds, which are in the order of a few hundred megabits per second, give or take [10]. We may or may not choose to optimize the dataflow, we may tinker with resolution, and we may want to be conservative and allow for higher-bandwidth digitally enhanced input data, but we are anyhow approaching practicality. VR is moving toward full immersion, and these capabilities are certainly approaching the capacities needed for sensory input communication and instantiation.

Upper bounds To communicate the complete brain state is of course something else entirely. In the book *The Singularity Is Near*, the author estimates that our total brain storage capacity is around 10^{18} bits. To convey a brain state for a particular emotion or otherwise it will not require this much capacity, but it is a useful measure for an upper bound. To transfer such amounts of data will be challenging, but this challenge is by itself not a fundamental challenge. That is, it is a grand engineering challenge, but it is in theory quite feasible.

9.3.4 HBC in the Extreme: Malleable Experiences and Artificial Memories

A captured memory or experience, an HBC object, may be modified and augmented much like a picture may be "photoshopped." And, ultimately, the HBC object may be entirely artificial in that it is created without ever being originally experienced by a human. To carry out the CCI sequence on a brain/mind level for a memory or experience would be very much akin to the visions depicted in the short story *We Can Remember It For You Wholesale* [16] that was mentioned earlier. The story is about artificial and constructed memories, which would be possible with Full HBC.

We shall return to the issue of HBC object integrity, but suffice to say that unless integrity is assured, a Full/Deep HBC human will have no way of knowing if the memories and experiences are his/her own, somebody else's, or even entirely fabricated. The basic problem manifests even with perfect HBC object integrity in place. If we choose to instantiate other entities' experiences and memories, or even fabricated ones, then it seems clear that this must necessarily change us. The more pervasive, encompassing, and compelling the HBC object is, the larger the impact will be.

A HBC experience may be *very* compelling, and HBC therefore has great capacity to influence us. This may be used by malevolent forces and it is not difficult to envision this as a tool of extreme oppression. Visions such as those in Orwell's dystopian novel *1984* [22] comes to mind, but they may not be the likeliest. Oppression may be through more subtle means, and HBC visions may have such compelling powers that one may perhaps fear futures akin to *Brave New World* [23] more. In Huxley's dystopia, the people were molded and conditioned to behave as the rulers wanted. Indeed, they mostly came to prefer the outcome. In the 1999 movie *The Matrix* [17], we are faced with a reality that is fully simulated and fed to the human brains directly. The artificial world is so compelling that the Cypher character, which was freed from the matrix, choose to return to the matrix. The people in *Brave New World* were conditioned since before they were born.

Neil Postman, in his classic *Amusing Ourselves to Death* [24], points to the fact that the media has become all about entertainment and commerce (advertisements). There is little or no content, but the form is very persuasive and captivating. One may indeed fear a future with pervasive HBC visions, where truth is so easily exchanged for entertainment. Of course, truth and reality will then be malleable concepts too.

We are, at least to some extent, the sum of our history. If that "history" is fully fluid, then it begs the question if we remain the same with different histories? That is, the "I" becomes an even more fluid concept than it already is. Combine this with the insights from *The Self Illusion* [1], and we might end up as "the sum of someone's histories" (Feynmanesque pun intended).

9.4 Brains and Minds

9.4.1 The Human Brain

Our brain is an organ that serves as a centralized control center of the nervous system. In a typical adult human, the brain weighs about 1300–1400 g and contains in the order of 80–90 billion neurons [25]. The cerebral cortex (the largest part) is estimated to contain about 30 billion neurons. The neurons form a densely coupled network, and each neuron is connected by synapses to several thousand other neurons. The neurons communicate by means of long protoplasmic fibers called axons. The axons transmit signals to other parts of the brain or body, targeting specific recipient cells.

Actuation The brain acts on the body in two main ways. Firstly, it induces muscle activity by transmitting signals to the muscles. The brain can thereby exert relatively rapid responses to external stimuli. Secondly, it initiates secretion of hormones. The impact of released hormones tends to be comparatively slow.

Sensory input We have already addressed this topic, but suffice to repeat that the input is in the order of a few hundred megabits per second. That is, however, the maximum peak capacity. The average input is much less than that.

9.4.2 The Brain as a Computer

The operations of individual brain cells are fairly well understood in isolation. However, with billions of cells and with a huge interconnection network, we are still only scratching the surface with respect to understanding the brain as a whole. What is very clear, however, is that the principal mode of operation in the brain is parallel processing and pattern recognition. For instance, in the neocortex almost all the neurons are organized in a highly repetitive way.

9.4.2.1 Computational Capacity

The neocortex is closely associated with conscious thought and cognition, and the size of this brain part is what really sets us apart from other mammals. The basic structure in the neocortex is the cortical column, with approximately 60,000 neurons. There are about half million of these columns, making for a total of about 30 billion neurons in the neocortex. Within the columns there are pattern recognition circuits. These consist roughly of 100 neurons and perform the basic recognizer function. Thus, we have about 300 million recognizers available. The estimated maximum computational capacity of the whole brain is in the order of 10^{15} calculations per second (cps).

9.4.2.2 Memory Capacity

We also have memory. The human memory system is less well understood, but it is obvious that we have one and that it has considerable capacity. The precision, speed, and accuracy of our memory system are not anywhere near the state of the art in computer memory systems. In terms of raw capacity potential, we have that the theoretical maximum memory capacity is in the order of 10^{18} bits. These capacities are surely impressive, but they are not outside the reach of what we may replicate in silicon. In fact, the raw capacities are within range today, and the progress in processing capacities is steadily increasing. That makes it clear that, at least in terms of raw capacity, it is soon feasible to emulate the human brain. Computer memories are also much faster than our brain memory system.

9.4.2.3 The Computational Brain

Kurzweil, in Chapter 3 in "How to Create a Mind: The Secret of Human Thought Revealed" [26] and Chapters 3–4 in "The Singularity Is Near: When Humans Transcend Biology" [10], provides descriptions of the human brain as a computer. Kurzweil's views are somewhat controversial and his way of communicating the ideas can be provocative. There are many counterargument and objections to be found, but, predictably, many of those are at least as emotionally charged as the views of Kurzweil himself. Kurzweil, coming from the AI field, is not shy of expressing his Strong AI views. However, Kurzweil is far from the only one considering and exploring these topics, and neuroscientist, are also very much engaged in this topic.

In the recent book *The Future of the Brain: Essays by the World's Leading Neuroscientists*, one explores the brain in many dimensions, including essays on mapping the brain, the brain–mind relationship, the computational aspects of the brain, and even the prospect of simulating the brain. In particular, in the essays "Whole Brain Simulation" by Hill and "Building a Behaving Brain" by Eliasmith, the authors are clearly seeing the brain as a computational unit that may be simulated/emulated. It is also noted that the book contains a part entirely devoted to expressing skeptic views. Still, these views seem to be more concerned with the immaturity of the field than with any fundamental objection to viewing the brain as a highly complex and highly adaptable biological computer. Others, like Bostrom in *Superintelligence* [27], clearly see the brain as a computer, but he also seems to think that whole brain emulation is further away in time (though not much further away).

There is therefore in reality little or no dispute over this topic, and while there may be real differences in opinion between Strong AI proponents, neuroscientists, and brain researchers, it seems that "in principle" they all agree that the brain is a computer and that it can be simulated/emulated.

9.4.3 The Mind

9.4.3.1 The Hive Mind Concept

Marvin Minsky, in the book *Society of Mind* [2], simply stated that the "the mind is what the brain do." The "society of mind" also alludes to the mind as consisting of many more or less autonomous parts. The mind may therefore very well be a hive mind at the lower cognitive levels. Consciousness and the self may then be viewed as emergent properties of a hive mind.

9.4.3.2 The Biological Premises

In the "The birth of the mind..." [28], the author explains in great detail the biological basis for the brain and how a mind emerges from a sufficiently capable brain. The author takes a holistic perspective while explaining how evolution, genetics, epigenetics, gene expression, cell biology, and neurobiology all play together. This does not axiomatically explain why there is a mind, but the explanations are compelling and, from a computational point of view, seem credible.

9.4.3.3 Consciousness and Self

The mind is closely associated with consciousness, and in the book *Consciousness and the Brain* [29], the author chooses this perspective. In a very condensed form, the main hypothesis is that "consciousness is a process of brain-wide information sharing." That is, it is a higher layer emergent characteristic that cannot easily, or entirely, be explained by the sum of its constituent part. Consciousness is also a prerequisite for recognizing our own "self." As strongly alluded to in Ref. [1], the self and the nature of identity is in some ways a fabrication, albeit a very useful one.

9.4.3.4 The Mind 2.0

We cursorily note that the "hive mind" idea may be extended further to encompass full-scale minds operating in concert to achieved higher-level cognition. This kind of super-mind may be the next logical level when one transfers minds to nonbiological computational substrates.

The SF book *Neuromancer* [30] prominently featured two AIs named Wintermute and Neuromancer. Wintermute was a hive mind and Neuromancer was a "personality" AI. Neuromancer could host other minds, and these were permitted to be dynamic. The fusion of Neuromancer and Wintermute was, however, not successful. Shortly after they fused, they achieved new levels of consciousness, but were unable to handle it and subsequently broke down into a set of lesser AI entities.

9.4.4 Technological Challenges in Context

There are tremendous technical challenges to achieve full immersion CCI of sensory data from a person to another. Given general technological progress that is already laid out in current roadmaps for processing power, communication capabilities, and sensor/actuator improvement, these challenges will be overcome. The timescale may be a decade or two, but in human history this is a very short time horizon.

To achieve brain-to-brain CCI is something quite different. We claim that there are no fundamental barriers, and the obstacles and challenges will be overcome. That is, we claim that what remains is in the domain of engineering and natural technological progress. In Ref. [10], the author is confident that "brain uploading" will be feasible within a decade or two. The argument is largely based on Kurzweil's "law of accelerating returns" [18], which roughly states that technological advances are used to advance technology. This, if taken seriously, is a prescription for exponential progress. When one considers Moore's law and similar phenomena, then certainly this view seems justified.

Breakthroughs, like practical quantum computing, seem likely to occur in this time frame [31]. Now, the time frame of a few decades actually matters little here. It is of little consequence if it takes a few decades or a century; the essential point is that human progress in brain research and ICT in general seems not to be heading for any fundamental limits, which implies that we are likely to get there. Hence, we postulate that brain-to-brain CCI capabilities will be developed.

If we take the idea that the brain is a biological computer seriously, then obviously we can create a simulation of this brain running on a different computer. That is, technologically it may be a very complex challenge, but in principle it is just another virtual machine instantiation project. The ultimate HBC object is the whole brain/mind, and it seems indeed theoretically feasible to apply the CCI sequence to this object.

The computational substrate capable of running a virtual brain will need to be very capable. Mind models that do not require a human brain (pure AI) could very well run natively on that substrate.

9.4.5 Virtual Brains

The electronic brain concept has several variations to it; among them is the possibility of fully emulating all functions of the biological brain. As mentioned earlier, the essays "Whole Brain Simulation" by Hill and "Building a Behaving Brain" by Eliasmith (both in Ref. [12]) point directly to this possibility. These essays are not just fictional essays, but rather point to research programs.

The essay by Hill is directly related to the HBP research program, which is financed by the European Commission. The project is an ambitious one and will no doubt advance the research frontier in this area. More information about HBP can be found on www.humanbrainproject.eu.

Eliasmith highlights the SyNAPSE project run by IBM and supported by DARPA. The goals are pragmatic and practical, including developing electronic neuromorphic machine technology. Ultimately, one wants to design and build a cognitive computer that replicates the functions and the architecture of the brain. These artificial brains may be used in robots or control systems, and it is imagined that the intelligence would scale with the size of the neural system. That is, one envisions increasing the intelligence by increasing the number of neurons and synapses and the connectivity.

It is clear that the SyNAPSE project is about building a virtualized brain. A virtualized brain may operate extremely much faster than our biological brains. That may make it an extremely efficient simpleton servant, but it can easily also be more than that. The *Terminator* movies, mentioned earlier [3], painted a rather dismal future in which Cyberdyne Systems created a chip that acquired intelligence and became sentient. The being, called Skynet, grew malevolent, and when its intelligence grew at a geometric rate, it became the most powerful adversary.

9.4.6 Types of Minds

In *Superintelligence* [27], the author distinguishes between different types of superintelligences. Among these are the following:

- Speed superintelligence minds
- Collective superintelligence (aka hive mind)
- Quality superintelligence (true AI)

These mind types can easily be mapped to Deep HBC. The following exposition is inspired by Bostrom's account (chapter 3 in Ref. [27]) but sometimes deviates slightly from it.

9.4.6.1 Speed Superintelligence

This type of mind can very well be a human mind, but vastly faster in every respect. In its weakest form, within the Full HBC paradigm, this type of speed superintelligence is still hosted in the human brain but with many artificial enhancements, including a fully digital CCI interface. This would enable learning to proceed at the full peak speed of our entire civilization, only impeded by the interfaces to the biological parts of the brain. Obviously, at a societal level, research progress at the frontier of knowledge would still be limited to biological level speed, albeit augmented by computers. The Deep HBC version is for the human mind running on a nonbiological computing substrate. All computation, all memory accesses, and all input/output are performed at extreme speeds. The relative speed difference between processing/thinking and input/output, including actuation and interaction with the physical world, would be a problem. Bostrom envisions that these minds would likely want to be located physically close to each other to reduce the inevitable communications delays. The speed of light becomes the bottleneck, and proximity becomes the only viable way to circumvent the problem.

It is not clear if such speed superintelligence would actually be smarter than us, but it would certainly outpace us. Since we do not fully know what intelligence is, it is not clear that faster automatically means smarter. "Would a dog be superintelligent if we ran the dog brain superfast?," one may ask. So, it may be that speed improvements by itself will have limits, at least unless we are above some critical intelligence threshold.

Concerning speeding up a human mind, this would enable it to achieve much more, and with Kurzweil's law of accelerating returns in mind, it would enable it to reach new levels at a very high speed. By itself this may not constitute an "intelligence explosion," but it would certainly be a most powerful tool toward further progress.

We observe that the identity of the mind does not at first seem challenged by the "superspeed" property. Technically, our memories may appear to be "the same." However, if we actually become smarter and more intelligent, then that would obviously change us and thereby the self-reflective "I."

9.4.6.2 Collective Superintelligence

Bostrom also envisions a collective superintelligence, where many autonomous smaller intellects team up to reach higher emergent cognitive levels. Some may argue that this is already happening with ubiquitous Internet access and collective wisdom, but it will of course be greatly amplified by minds communicating directly. One may envision this within the Full HBC paradigm. It would not principally be all that different from today, but learning would be very fast, and all communication between the minds would be orders of magnitude faster than today.

However, it is within the Deep HBC paradigm that collective superintelligence really seems to make sense. Each lesser intellect may be a speed superintelligence, and together they may be operating as more or less tightly coupled minds. While it may be that none of the constituent parts of the hive mind really becomes superintelligent, the emergent hive mind may very well have higher-level cognitive powers.

The identity of the lesser intellects may or may not be challenged in this scenario. The lesser intellects will appear to be identifiable and may exist more or less oblivious to the fact that there are an emergent intelligence/ mind present. Whether or not the integrity and/or privacy of the lesser intellects are threatened is difficult to tell, but there certainly is a risk. As seen from the emergent mind, the threat may be exactly that the lesser minds have too much independence. However, the collective mind will not likely critically depend on any individual constituent intellect, but it may fear impacts that quickly spread among the lesser intellect. A lesser intellect mind-virus would spread at breakneck speed and would have the power to disrupt the emergent mind.

With respect to identifying the emergent mind, it seems certain that this will be possible. However, the emergent mind may not be a fixed entity. It could be a transient being existing for only a brief period and perhaps also changing so fast that it is meaningless to assume it to continually be the same entity.

9.4.6.3 Qualitative Superintelligence

Bostrom also defines a qualitative superintelligence. This type of mind should be at least as fast as a human, but quantitatively vastly more intelligent. That is, this is an AI that do not require a brain-like computational substrate or that mimics the structures of a human mind. From an HBC perspective, this type of mind is outside the scope since there are no "human" aspects present at all.

The issues and scenarios concerning identity, integrity, and privacy for an AI may be similar to that of the emergent collective mind. One may object that privacy may not mean much to an AI, but then again it may not mean much to an emergent hive mind either. At least, we should not expect privacy to have any humanlike meaning anymore.

9.5 Security and Privacy for HBC

The following is an initial outline of a security and privacy analysis for HBC. There is no real attempt at completeness, but rather to highlight the most important aspects. The assets are the HBC objects and the main goal is to see what can be done to secure CCI sequences. The HBC level coverage encompasses both Weak HBC and Full HBC. Deep HBC is also addressed, but this coverage is mostly based on educated guesswork.

9.5.1 Uniqueness and Identification

9.5.1.1 Entity Identification

There is commonly a need to uniquely identify an entity. The entities in question are processes that act autonomously or they are people. They may also be "legal" object like a business.

Whatever the case, we often need to establish the identity of an entity. Philosophically, an identity is often said to be the essential intrinsic property of a thing that defines it. It applies to the sameness of an object (over time) and it must be a distinguishing trait. In practice, we invent different types of references or identifiers that we use. These are practical short hand ways of indicating the entity. This includes names, addresses, social security numbers, email address, object identifiers, and so on.

The concept of identifiers can be extended to group identifiers. The uniqueness property would then apply to the group. Identifiers may also refer to anonymous identifiers (a paradoxical term for sure).

9.5.1.2 Object Identification

The HBC objects will also need to be identified. There therefore needs to be some sort of object identifier. There is in general no shortage of identifier schemes, and to create one for HBC objects is also easily done. The thing to note, however, is whether the identifier is intended to identify the type of object or if it is intended to identify a specific object.

For digital object, which is easily copied, this makes a big difference. We note that identifiers that are intended to refer to a specific object are more complex. These identifiers are also harder to verify.

9.5.1.3 Entity Uniqueness

Identifiers, like the original bar codes, are not intended to uniquely identify a single object. Rather, these identify the type of object.

For identifiers that identify specific objects, there is a requirement that these identifiers are unique. That is, it is obviously possible to copy a digital object A, but then the copies of A (A', A",...) are technically the same object.

An HBC experience, captured in an HBC object, can be copied and shared by many minds. For IPR reasons one may want to know how many of these objects have been consumed, and so there may be a need to be able to track individual copies, presuming, of course, that IPR issues make any sense in this context.

The HBC object to be identified may potentially be an entire mind. The captured HBC object, a mind, is then an HBC entity. An AI would be in a similar situation. It seems reasonable to assume that these entities should have unique object identifiers. However, for Deep HBC it is clear that it will be fully possible to instantiate the same mind multiple times. This problem can of course be resolved, and it is quite possible to have one type of identifier for a captured mind and another type of identifier for instantiated minds. Then, uniqueness can be had for instantiated minds in much the same manner as one may run multiple instances of the same program on a computer. Each instance is represented by a unique process, which has its own unique process identifier.

It will therefore be a firm assumption that all instantiated entities will have unique entity identifiers. The question is much more complex for emergent hive mind-like entities, but sentience and consciousness seem to be oriented around identity, and so it is presumed that this type of entity will also have a unique identity. Thus, there will be identifiers for this type of entity too.

9.5.2 Creation, Ownership, Existence, and Death

This is a peculiar subsection to write. We have tried to only analyze the possibilities and avoid moral and ethical issues.

9.5.2.1 Creation and Ownership

To "create" a new entity would in principle be no different than creating a new process on a computer. One simply instantiates the new mind. The act of instantiation may require rights and access to computational resources, like a virtual brain or similar. But, otherwise, it should not be difficult to instantiate a mind in a Deep HBC reality. The rights in question must be enforceable, and the methods and schemes for the enforcement may be based on cryptography and security protocols.

Another aspect is that of ownership. Ownership seems straightforward for HBC objects that merely consist of captured experiences, but it is not clear how it should scale for HBC objects that encode minds. To own an encoded mind object is one thing, as it in principle is just a digital object. However, an instantiated mind is something different, and the notion of ownership becomes more difficult to handle. That is, technically it does not have to be difficult to handle this aspect, but it raises all sorts of ethical questions. Will we tolerate slave minds? Can such an entity have free will and is it entitled to freedom?

Ownership does translate into authority, as in jurisdiction over an object. Ultimately, this authority is total, but it is not at all clear that HBC entities will be handled this way. It may very well be that there will be higher "societal" authorities that retain this power. Technically, one may envision an authority system regulated in similar ways as in the case for our society, where there are law-and-order authorities. These authorities will need ways to enforce their authority.

9.5.2.2 Existence and Death

For minds running on nonbiological substrates, there will be no ordinary aging processes. The mind may run on a virtual brain or some other computing machine model. Such a mind may migrate to a different hardware should the need arise, and existence may in principle be eternal.

It is not obvious that eternal existence is wise or desirable. Another question is that of resources. Computation does require resources and it may be the case that an entity runs out of resources. Or the creator decides that the entity is no longer useful. Processes on a computer may be terminated/killed, and so it is similarly possible to terminate instantiated minds. Again, in the analogue case of a present-day computer, to terminate a process requires rights, and so it would seem natural for the termination of a mind to require rights. In the computer analogy, we note that a parent process will normally be allowed to kill a child process. However, killing people is not normally permitted and so the validity of the analogy is limited, and it may be completely invalid for instantiated HBC entities. But, of course, people do not live forever, and in a situation where there may be competition for resources, it is reasonable to assume that entities will indeed need to be terminated.

Thus, we may assume that HBC entities will not exist forever. This implies that there will be some kind of "death" or termination for HBC entities. Just as for people, there will certainly be a need for protection from illicit termination. There may also be a need for some sort of protocol for lawful termination.

9.5.2.3 Apoptosis

Apoptosis is a genetically directed process of cell self-destruction. It is also called "programmed cell death." Among fears of runaway AI entities, there have been papers concerning safety schemes. In this vein, apoptosis, applied to electronics on the chip level, has been proposed as a way to have "programmed AI death" [32]. This would, it is claimed, aid in preventing unwanted intelligence explosions and would also serve as a way to handle death by aging for AI entities.

One may wonder if an HBC entity, or an AI entity, would accept to be terminated this way or if it would direct its resources into migrating to new hardware well before the current hardware would "die."

9.5.3 Load and Store Minds

9.5.3.1 Processing and Timeliness

Assume now that an instantiated HBC entity exists. What kind of rights to processing does this entity have? Humans are born with a physical brain that thinks at its own speed, while others cannot indiscriminately impose limits to that speed. But what about an HBC entity? Can such entity be regulated to run at certain speeds only? Can it be stored and retrieved, paused, and muted by external authorities?

As humans, we do not get free access to resources. People actually starve, and they even die from lack of resources. An HBC entity may very well be in a similar situation. Self-preservation is a strong drive, and a starving HBC mind would naturally try to improve its conditions. Ambitious HBC entities may also want to expand and become even more intelligent. In fact, this may be a likely scenario.

Technically, it will likely be possible for an HBC authority to exert "processing" control over other HBC entities. And unless there are infinite resources, it may even be necessary for HBC authorities to exert this type of control. The controlled entity and the authority may have very different objectives and goals, and this may be the source of conflicts. Authority is therefore dependent on power and availability of methods for enforcement.

9.5.4 How Many of You Are There?

It goes without saying, but an HBC entity can be instantiated multiple times. These instantiations may overlap in time; they may even meet or team up to become a higher-level version of the mind.

Technically, it may not be too difficult to differentiate between the different instances of the same mind. Just as identical twins are very similar, the HBC twin minds will be very similar at least initially, and this is also analogous to identical human twins. Different exposure and experiences will lead to different genetic expressions, and over time the difference may be pronounced. One may assume that this applies to minds as well, and clearly they will be formed by the information they process and the internal structure they develop as part of that processing (the learning).

Perhaps one may view "blank" HBC minds as a kind of "mind stem cell." These "blank" minds would be malleable, and they can potentially fill many different roles. As part of a hive-mind, this could make a lot of sense, and these constituent minds may perhaps even have to be from the same template mind to develop peacefully. In this sense, a higher-level "I" of this super-mind may still, somehow, be you.

9.5.5 Principal Entities

We need to identify the principal entities involved. By entity we here mean legal entity, and we presume that the entity is autonomous. This may be a person, a business, or, in our case, an HBC-enabled entity. The principal entities are all recognized entities, participating in some kind of transaction.

For the purpose of this paper, the transaction in question will be the exchange of HBC objects. We denote the principal entities as entity $\in \{A, B, C,..., Z, A1, B1,...\}$. The size of the set may grow large.

9.5.6 Disinterested Parties, Authorities, and Intruders

There may be entities involved that do not directly participate in the transactions involving the principal entities. These parties are disinterested in the sense that while they may perform some service, their interests are limited to this.

An authority is any entity that has powers to arbitrate between entities or otherwise to regulate the behaviors of the principal entities. This could be according to laws and regulations, but any powerful entity that may interfere with the interactions between entities are an authority by definition. An authority may also be a disinterested party.

Additionally, there will be adversaries involved. This may be a dishonest principal entity or some other entity that will attempt to act in ways that are detrimental to the interests of the other principal entities. We generally use the term **intruder** for these types of actors.

We denote the disinterested parties and authorities in a similar vein as for the principal entities, if we denote them at all. In terms of security modeling, the intruder is an abstraction and we therefore have only one intruder.

9.5.7 Actors

The actors are the collection of principal entities, disinterested parties, authorities, and the intruder. The actors may have different rights and authority.

9.5.8 Assets

Assets are the valuables. These may be material or immaterial. For the purpose of this paper, the assets are the HBC objects. We denote these objects as *asset* $\in \{a, b, ..., n\}$. Observe that the different actors may attribute different value to the assets. Note also that, in the extreme, an HBC object includes a fully captured mind. When such an object is instantiated, it becomes an entity.

The rights of such entity is a philosophically interesting question. The entity will be a "child" entity, and it may only be allowed transient existence.

9.5.9 Trust, Trust Relationships, and Dependence

Trust is typically defined as reliance on the integrity and ability of an entity to behave in accordance with expectations. That is, trust may be viewed as beliefs about the future behavior of another entity. One sometimes differentiate between trust and trustworthiness, thereby differentiating between an intention to act as expected according to protocol and the ability to actually do so.

Trust relationships defined the trust that an entity has in another entity. In principle these relationships are directional and conditional. The fact that entity A trusts entity B does not imply that the reverse is also true. Trust may be mutual, of course, but asymmetries are to be expected. The asymmetries may be multidimensional, but an obvious aspect is a power/control/authority asymmetry. There may be conditional constraints on trust with respect to time, location, amount, and so on. Trust may be transitive, derived on the basis of other trust relationships. Note that the transitivity may be conditional.

Additionally, we note that dependence plays a role with respect to trust. For instance, if entity A is totally dependent on entity B, we have a situation where trust may not be justified as such but where dependence dictates trust. That is, A has no choice but to trust B. We note that dependence is normally associated with asymmetries in power and influence.

Also, a prerequisite for trust in an entity is that the entity can be identified and that the identity can be verified (entity authentication). The "authenticated" state must be maintained in a credible way for the duration of whatever trusted transaction that is commencing. Finally, we observe that an entity may itself be an asset.

9.5.10 Threats

A threat is often defined as an intention or determination of an adversary to carry out some mischievous act toward an asset. The motivation behind a threat is not necessarily important, suffice that there be an intention behind. One may keep morality and ethics aspects out of the question; all that is needed is that an intruder has the intention to do something with or to an asset a that some other entity A objects to. As seen from A, the intruder poses a threat to A's rights concerning asset a.

Normally, one is quite specific about the nature of the threat. For instance, the threat may be directed toward the confidentiality of the asset, the integrity of the asset, and so on. One may distinguish between so-called advanced persistent threats (APT) and more mundane or opportunistic threats.

9.5.11 Vulnerabilities and Exposure

Vulnerabilities and exposure are related to the assets. Specifically, the vulnerabilities refer to the protection of the assets. With no protection, the assets are highly vulnerable. With strong protection and no known flaws in the protection, there will be few, if any, vulnerabilities.

However, the fact that an asset is highly vulnerable does not mean that any threat necessarily will be perpetrated. Exposure plays a vital role here. If the asset, or the vulnerabilities, is not exposed (accessible) to other parties, then there may be no exploitation.

The exposure in question is also relative to knowledge about the exposure. The would-be intruder may not know about the asset, the vulnerability, or the exposure. Also, there may be parties that are so-called opportunistic intruders. These will not engage in sophisticated persistent attacks or anything even remotely like that, but they may, willfully, transgress rights and attempt illegal access to an asset if the opportunity presents itself.

9.5.12 Attacks

Attacks are threats that are instantiated by a threat actor. For an attack to be successful, the asset must be exposed and it must be vulnerable. Note, again, that "success" here is relational and relative to the entities involved and the rights that they may perceive to be violated.

Since perspectives differ, what actually constitutes an attack may depend on the point of view. That is, where one HBC entity sees an attack, another entity simply sees a case of resource allocation.

9.5.13 Security Requirements and Services

For a digital object, like an HBC object, we may have several different security requirements. We may want assurance about:

- The identity of the originator of the HBC object
- The identity of the owner of the HBC object
- The integrity/authenticity of the HBC object itself
- The secrecy/confidentiality of the HBC object

Observe that the secrecy/confidentiality is a relational property, which depends on who one wants to see/read the object contents. For privacy, one may also want to hide the fact that a certain party acquired certain assets. Again, the requirement may be relational, and it may matter a great deal who may or may not know these facts. The requirements need to addressed, and here we have the security services. For instance, encryption may facilitate confidentiality.

9.5.14 Security Architecture

A security architecture is set of security schemes and measures that, hopefully, address all security requirements in a consistent and complete manner. A security architecture is often domain specific but should provide generic services within the intended domain. It is utterly premature to try to define a security architecture for Full HBC and Deep HBC, but it may be prudent to initiate work to define a Weak HBC security architecture.

9.5.15   Trust Basis and Credentials

What is real and what is emulated? If an entity can emulate something perfectly, then a copy may be indistinguishable from the real thing. For security, and cryptography in particular, we use different types of secret credentials. All security hinges on these objects, keys or key material, being secret. For the purpose of hiding these secrets from other entities, we normally choose to rely on dedicated trusted hardware. The hardware then, it is assumed, has physical integrity, which effectively prevents any but the owner from accessing it.

This is all very well, but with virtualization technologies, how can one tell if one is accessing the trusted hardware? Imagine, for instance, that you use your smartphone to carry out some transaction on the Internet. While you may superficially perceive that it is you who carry out the transaction, it is quite clear that there will be some software that acts as your proxy. That piece of software may need to authenticate itself as acting on your behavior, and to do so it may need to access a secret authentication key. This key, it is expected, is stored in secrecy in some trusted hardware, say the SIM card. However, how will the proxy software know if it has been deceived? The software cannot in any meaningful way have assurance that it actually accesses the trusted hardware.

These problems are not unique to the HBC scenarios but apply to almost all modern-day scenarios. The whole cloud computing paradigm is centered around virtual machine technologies, and HBC systems will face the same problems. This is not to say that HBC systems will be cloud based, even though one may envision that too.

9.5.16   Security Goals

We commonly have a set of security goals that we want to achieve. These include, but are not limited to, the following:

Entity authentication This basically means the process to corroborate a claimed identity. For Weak HBC, one needs to embed existing personal identifiers and credentials into the security architecture. Strong, but otherwise normal cryptographic methods should be sufficient.

For Full HBC and Deep HBC, this will be akin to verify the entity identifier of an instantiated mind. It's hard to predict the exact kind of methods to be used but, there is an absolute need for a way to ensure the integrity of the process (the authentication procedure). Confidentiality will likely also be needed. This pertains to the credentials.

Data confidentiality This pertains to secrecy of the object and concerns the ability to protect an object from being seen/read by any unauthorized party. It may refer to data (payload) or it may refer to metadata (identifiers, type-of-service, address information, port numbers, and so on). Strong confidentiality is a prerequisite for privacy, whatever that may mean in an HBC context. It is hard to envision the ability to keep data confidential without some kind of bottom-layer (hardware) assurance.

Data integrity Data integrity refers to the accuracy and consistency of the data. In terms of security, it refers to projection against authorized modification (write) of the whole data element or parts of it. This may include deletion. The protection is commonly for detection of unauthorized modification and does not normally cover protection against modifications *per se.* That is, there will be a cryptographic checksum of the object that cannot be forged by any unauthorized party. The checksum accompanies the object and is the proof of the integrity. Strong integrity is a prerequisite for assurance. It is hard to envision the ability to maintain integrity without some kind of bottom-layer (hardware) assurance.

Authorization Authorization concerns access rights. An authorized party shall be able to exercise his/her rights. There must necessarily be a granting authority, which administers and/or owns the resource/object. The set of authorizations amounts to the security policy.

Accountability This property is to ensure that an entity performing an action is accountable and responsible for that action. This means keeping track of all relevant events and being able to use the log later to hold the entities responsible. We may refer to security logs or secure audit trails.

Availability It is vital to ensure that the resources or objects are available to the authorized parties as indicated in the security policies and service policies. It is a complex trait, and the security part is normally only about to prevent or mitigate so-called denial-of-service (DoS) attacks. Availability, in the context of instantiated HBC minds, does take on a special meaning.

Security architecture The set of goals and the circumstances, together with the security policies, are what defines the security architecture. To achieve the goals one needs to deploy a set of security protocols and security schemes, which again builds on cryptographic primitives, cryptographic key material, trusted modules, and so on.

9.5.17 Security Primitives and Absolutes

Security primitives are the lowest-level of tools. They provide a basic function, upon which other schemes and protocol is built. The lowest-level ones are the cryptographic primitives. They are based on mathematical constructs and provide certain features, which permit the security primitives' authentication, encryption, and others, to be derived from them.

9.5.17.1 Computationally Infeasible
A common mathematical property one seeks to incorporate is that of intractability. That is, one wants some functions to be computationally

infeasible to run, unless one has some secret information. Generally speaking the complexity is associated with the block size or key length of the cryptographic primitive. An N bit key should have close to 2^N entropy, and extending the key length would thus increase the complexity in an exponential manner. Provided that complexity actually rises exponentially with the field size, a strategy of extending the keys will continue to work for conventional computers.

9.5.17.2 Exponential Complexity and Quantum Computation

We just cursorily mention that there exists no proof that any of the cryptographic functions actually have exponential complexity with respect to breaking them. Also, certain kinds of computers may actually be able to parallelize its operations and seek out a huge number of possible solutions simultaneously. There have been experiments with biological computes, but undoubtedly the so-called quantum computers are the most promising so far. A quantum particle can be in more than one possible state simultaneously (quantum superposition), like Schrödinger cat (being both alive and dead simultaneously). For entangled states, a quantum computer may in principle be in any of the possible states. The entanglement "problem" (as it was seen) was famously described in a 1935 paper by Albert Einstein, Boris Podolsky, and Nathan Rosen (EPR) [33]. The Schrödinger's cat "paradox" [34] is also known as the EPR paradox.

Today, the use of quantum effects to produce quantum computer is well under way. There are many theoretical and practical problems to be resolved, and a practical quantum computer may yet be years ahead, but suffice to say that if a practical quantum computer is produced, then the complexity of exponential problems may be illusory. Especially, almost all asymmetric cryptography will be fundamentally shaken and may be left utterly useless.

9.5.17.3 Trusted Base

Another kind of security absolutes is to have some cryptographic function and corresponding security credentials execute and reside on a piece of "trusted" hardware. However, is trust in hardware actually warranted? One essential property for the trust is that some parts are unchangeable and invariant by nature. That is, the hardware cannot be illicitly modified but remains as it was designed to be, somewhat similar in kind to a mathematical invariant.

However, is that actually the case for modern hardware? Today, it is increasingly hard to know for sure if a certain piece of hardware is hardwired or if it is reprogrammable or replaceable by some other piece of circuitry. When even the minutest chips may have millions of transistors, how will you actually know how the hardware works? What is virtualized and what is not? The problem runs deep, and the author investigated this and other trust-related issues in Ref. [35]. That paper was again inspired by Thompson's seminal 1984 paper "Reflections on Trusting Trust" [36], a very short and still timely paper that should serve well to shake your trust in software.

9.5.18 Privacy

The right to privacy has a deep and philosophical side to it. These aspects are not new but are becoming substantially more acute in an all digital world where the rights to personal privacy can so easily be violated.

9.5.18.1 The Right to Privacy

In the paper "The Right to Privacy" [37], the authors give a broad and compelling range of arguments for man's right to personal privacy. The authors wrote from a perspective of generic legal rights. The article was published in 1890 and clearly did not take into account digital aspects of personal privacy. However, the authors foresaw the need to capture privacy as a generic right and declared that the right applies to any modern device that can infringe on the personal privacy rights.

9.5.18.2 Societal Considerations

Privacy also has emergent dimensions that apply to society more than to individuals. These dimensions arise from the rights to free speech and similar rights, where part of the freedom is not to be persecuted for your views. This sometimes requires guarantees for anonymity. Privacy as a right is molded by the society and is often at odds with safety. Thus, privacy may be traded for more safety and less crime.

In the 2002 paper "The New Panopticon: The Internet Viewed as a Structure of Social Control" [38], the authors address the societal impacts of mass surveillance. The notion of a set of lesser Big Brother's is touched upon in Ref. [39] too.

All well-functioning nations will have some sort of surveillance scheme in operation. If for no other reason, they will at least need it for crime prevention and similar services. Of course, ordinary "lawful interception" tools can be abused, but in either cases we have that the capability for surveillance is there. The "lawful" Big Brothers may be recognized as benign, but this is very much a matter of trust in the authorities [40]. Of course, the ultimate Big Brother vision is captured in the book 1984 by Orwell [22]. Here, surveillance is very clearly a tool for suppression and control.

9.5.18.3 Control and Power

Since privacy is so closely associated with control, it is no surprise that there will be actors that want that control for themselves. Powerful authorities, like a national security agency, will obviously be more powerful with mass surveillance tools, as has been amply demonstrated by the National Security Agency (NSA). But, while the motivation may initially be credible and benign, it is very easy to use those same tools for oppression, illegal spying, and harassment of political opponents. Greenwald provides a fascinating account of the Edward Snowden story and the vast mass surveillance carried out by the NSA [41]. He also addresses many of the societal concerns one may have with lost privacy. The NSA mass surveillance programs are reminiscent of the base idea in the *Panopticon or the Inspection House* [42], where there is optical mass surveillance such that no one can be sure of their privacy.

9.5.18.4 Deep Privacy Questions

Privacy has many aspects. There are obvious technological ones [43], but the issue is deeper and more profound. Even today, there is almost no distinction between normal life and life in cyberspace, and privacy seems to be losing out. For an HBC reality, these issues will be pronounced.

Can there be freedom when the (Cyber) Big Brother [22] literally can see all? The panopticon vision [42] is an early example of a Big Brother that *may* watch you at all times. Can there be democracy without privacy? And what does the concept of democracy actually mean in a Deep HBC reality, where one may instantiate millions of copies of the same basic HBC entity?

Without privacy, can there be trust? This is discussed in depth in Ref. [44], and it seems that privacy is a very important component of trust. In "Reflections on Trusting Trust" [36] by Ken Thompson, we get a classic demonstration of how difficult it can be to trust anything digital. There are different layers of truths, but unless you know you have full control of all layers, you will never know if you have privacy. In these days of cloud computing and virtual machines, we clearly see that Ken Thompson's demonstration has deep implication. Literally, we can't know the base case (in the recursive layers) or how deep the rabbit hole goes [45]. Needless to say, but in an HBC reality, trust will be even more important and probably also harder to ascertain.

9.5.18.5 Privacy in an HBC Context

Technically speaking, privacy is very much about using security mechanisms to ensure that people, and objects associated with people, cannot be tracked. The tracking mostly concerns the so-called metadata, but obviously the data can by themselves also be privacy sensitive. In either case, one must ensure that nobody but the authorized parties can see the private data, will know about the fact that the private data was exchanged, and so on. We can then differentiate privacy in to several subcategories, including [46] the following:

- Data privacy
- Identity privacy
- Location privacy
- Movement privacy
- Transaction privacy
- Data control (all HBC objects)

A comprehensive technical account of privacy can be found in the book [43], which also contains many references to other works in this area. Still, the privacy challenges for HBC will be massive, and today's tools are almost certainly not up to the task. The urgency for HBC privacy is dire and this is the most important challenge. The consequences are severe even for Weak HBC, and they only grow more pronounced as we move along towards Full HBC and then further toward Deep HBC.

9.6 The Adversary

9.6.1 An Intruder Model for HBC

The classic Dolev–Yao intruder (DYI) [47] is an extremely capable intruder. It can:

- Read any message transmitted.
- Modify any message transmitted.
- Delete any message transmitted.
- Inject any message at will.
- Store every message ever transmitted.
- Utilize all information it has optimally.

The DYI has some formidable powers in that it can use all information ever recorded optimally to break cryptographic protocols. This is an extreme capability. Surprisingly, the DYI cannot actually break any cryptographic primitives. That is, the standard assumption is that cryptographic primitives are effective. Also, given that we are dealing with an abstract intruder model, there are no worries concerning short keys. One simply expects the keys to be "long enough." In contrast to real-world intruder, the DYI does not have any powers to corrupt a principal entity or his/her security credentials.

9.6.2 A Weak HBC Intruder

For HBC we have different levels, and thereby we may have different levels of HBC intruders. The first level, Weak HBC, is not so far from today's technologies. One may *choose* to have trust in specific hardware modules that perform the critical security functions. In this kind of scenario, it may make sense to model Weak HBC with a classical DYI.

One may of course extend the intruder to be able to modify the *capture* and *instantiation* parts. Here one is essentially left with the same situation we have today: hardware and trusted modules may be corrupted. We have no effective way of dealing with this, except reverting to trust and to higher authorities to punish those that cheat and corrupt. This may work, to an extent, for a large class of adversaries, but obviously not if the adversaries we have are the very authorities that we depend on for recourse to the law. We may also be worried that the authorities are poor at detecting fraud and that they are lenient with persecution and punishments.

9.6.3 A Full HBC Intruder

The situation is a much more serious one for Full HBC. The augmented human may not be able control the Full HBC equipment. For Weak HBC, this may be some special suit to wear or something similar, but if the HBC is integrated directly with our brains, the impact and vulnerabilities are very different.

To some extent there are no fundamentally new threats to Full HBC, but the risks are obviously graver and more acute. It is hard to estimate the probability of serious incidents, but the impact potential is clearly higher.

9.6.4 A Deep HBC Intruder

A Deep HBC intruder will be a very dangerous adversary. For Deep HBC, the entities are run on some computing substrate, but what exactly is the nature of this substrate?

We have already mentioned that the substrate may host virtual brains. Given this possibility, the human-mind HBC entity may be fully virtualized. An instantiated mind may not have any means whatsoever to decide if it is running on a bottom level. There may be a universal HBC virtual brain somewhere, in the same manner as for a universal Turing machine. Still, with perfect emulation there would be no way for the universal machine to know if it is not emulated itself. This is a fundamental problem.

In fact, it may be the case that no entity can be assured that it runs on the lowest level. Since that is the case, no entity can have full assurance that it is in control. Nevertheless, some level will be the lowest level and some entity may actually be in control. If that entity is aware of the fact is another matter. The ultimate Deep HBC intruder would be running on level $N-1$, where the target is running on level N (or higher) and may be hosted by the intruder. In such scenario, the Deep HBC intruder will essentially be in full control and may appear to have godlike powers.

9.7 Impacts

9.7.1 Threats at Large

An HBC reality will be disruptive in many ways. It has the power to irreversibly and fundamentally change our society. That, by itself, is a threat to the existing structures and powers-that-be. Even the mundane Weak HBC will likely have a profound impact on society. It will change how we perceive reality, and with VR we might even create our own reality. This will fundamentally change the ways we interact with other people and thereby change the modus operandi of our society. In some sense, we are already heading toward this change. The opportunities are tremendous, but the dangers and risks are also there in full force.

9.7.2 Singularity Visions and Existential Threats

The *singularity* is broadly expressed as a point in history when our intelligence will be transformed by technology and become millions of times more powerful than it is today. At this point the self-improvement cycle will be so fast that we will have an *intelligence explosion*. The key drivers are seen to be nanotechnology, genetic engineering, and AI. AI might be complementing our biological intelligence, or it might be self-sustained.

The so-called singularity visions are not associated with HBC *per se*, but for Full HBC and Deep HBC, it is clear that it will share common ground with the singularity visions. Even Weak HBC will have some similarities and in particular with the "accelerating returns" prequel to the singularity.

9.7.2.1 Intelligence Explosion

The singularity viewpoints were perhaps first expressed clearly and succinctly by Good in his 1965 paper "Speculations Concerning the First Ultra-intelligent Machine" [9]. Good was quite concrete on a number of aspects of the hardware and type of algorithms he foresaw as essential to constructing ultra-intelligent machines. He also had a few concerns considering the impact of these machines.

> Since the design of machines is one of these intellectual activities, an ultra-intelligent machine could design even better machines; there would then unquestionably be an "intelligence explosion," and the intelligence of man would be left far behind (see for example refs. [1, 8, 28]). Thus the first ultra-intelligent machine is the last invention that man need ever make, provided that the machine is docile enough to tell us how to keep it under control.
>
> —From Section 2 in Ref. [9]

Good does not use the term *singularity*, but obviously the *intelligence explosion* does represent a singularity type of event. He also clearly outlined the idea of self-improvement with ever more capable means.

Finally, Good included the conditional clause "provided that the machine is docile enough to tell us how to keep it under control." That is a fundamental aspect. The problem with a singularity, by the very definition, is that one cannot really know much about what happens after the singularity has occurred.

9.7.2.2 The Coming Technological Singularity

The 1993 opinion essay "The Coming Technological Singularity: How to Survive in the Post-human Era" [48] by SF author and mathematics professor Vernor Vinge is famous for explicitly introducing the singularity meme. Vinge considers the singularity to be inevitable and also sees great dangers in the transition. Vinge also postulates that the singularity will potentially permit us to live forever and that this by itself will induce us to want to become more intelligent (intelligence amplification). Vinge is decidedly skeptical but also sees great potential.

The intelligence amplification part is relevant to HBC, and while it seems to be most relevant to Full HBC and Deep HBC, it will also apply to Weak HBC (augmented on-demand memories, increased learning capabilities, and so on).

9.7.2.3 The Singularity Is Near

Ray Kurzweil is a futurologist and AI researcher. His book *The Singularity Is Near: When Humans Transcend Biology* [10] is his take on the singularity vision, and by now this vision is perhaps what most people associate with the singularity vision. This includes his accelerating returns theory and his postulates for improvements in genetic engineering, nanotechnology, and robotics (AI). Kurzweil is quite positive on his outlooks for the singularity, and he is also quite concrete. The singularity is to occur by 2045 and it will by and large be for the better of human kind.

Our HBC levels could very well be mapped onto the various stages in the process toward the singularity.

While developments toward a singularity is not written in stone or completely given, this author concurs with Kurzweil in the general view that we will approach a technological singularity if our developments proceed as they do now. However, there are also many events that may disrupt the progress. The risks associated with an ever more technology-dependent society are largely grossed over by Kurzweil. As seen from a security professional's point of view, Kurzweil technology optimism seems a little misplaced and even a bit naive. He could be right, but can we afford a situation where the singularity has arrived and it turns out that the future doesn't need us?

9.7.2.4 Why the Future Doesn't Need Us

In the essay "Why the Future Doesn't Need Us" [4], Bill Joy speculates that the intelligence explosion will be a bad thing. He seems to take the view that there is no inherent way of keeping an ultra-intelligent machine docile and that it also seems quite plausible that we are on the road to actually constructing ultra-intelligent machines.

9.7.2.5 Malevolent AI

In the 1946 SF short story *A Logic Named Joe* [49], we are faced by "a logic" that is much reminiscent of an extended version of Apple's Siri or Microsoft's

Cortana. In short, this "Joe" is an "intelligent" assistant much like those that we have accessible in our mobile phones. Joe was not malevolent *per se*, but he took the quest for efficiency too literally. Due to a minor deviation in his circuitry, Joe was able to bypass security filters and really answered all the questions that people had. Joe was described as ambitious, and he also took over other "logics" to achieve his goal. This caused total chaos, since people asked all sorts of inappropriate questions and Joe answered them all, ranging from how to carry out robbery, infidelity, and murder. In the end, Joe was disabled, but the technician that disabled Joe was obviously tempted to turn Joe on again to reap some of the benefits that Joe could provide. We note that in an HBC reality, personal assistants might be directly integrated with our brains. If one is subverted or if we can subvert it, then obviously we may use the assistant for crime. Also, we could very easily be fooled by a malicious Joe.

A much more malevolent AI is described in Harlan Ellisons' *I Have No Mouth, and I Must Scream* [50]. The AM computer, an intelligent computer designed for warfare, has not only killed off its competitor computers, but it has also killed off most humans. AM hates people, and the remaining few people have been tormented and tortured by AM. The computer keeps these people alive just to torment them, and ultimately it turns the last person into a blob of flesh without limbs so that he cannot kill himself—a dark vision to say the least.

9.7.3 Why System Fails

9.7.3.1 Normal Accidents
The 1999 book *Normal Accidents* [51] by Perrow is not about AI or singularity views at all. It is, however, about complexities and tightly coupled systems. Such systems are hard to analyze and understand. When they fail, and they invariably seem to fail, they fail in spectacular and totally unexpected ways. Perrow identifies three basic conditions that make a system likely to be susceptible to these types of failures. These are the following:

- The system is complex.
- The system is tightly coupled.
- The system has catastrophic potential.

These conditions cover basically any modern ICT-enhanced critical infrastructure system. We can safely expect these systems to be augmented by narrow AI systems on a regular basis. It seems highly likely that any AI systems, or HBC system for that matter, would be susceptible to normal accidents.

9.7.3.2 Black Swans
Before the Europeans knew about Australia, there were, to them, only white swans. A black swan was not even conceivable. In retrospect, a black swan is easily explained and no one is surprised by its existence.

In the book *The Black Swan: The Impact of the Highly Improbable* [52] by Taleb, we have a complementary view to that of Perrow. The main thesis in this book is that we are generally unable to estimate risks from novel and highly improbable events. This has to do both with misunderstandings concerning statistics for highly improbable events, where there generally are too few events to reliably establish the distribution, and with our psychology and our inherent inability to evaluate novel risks. The Taleb book is also concerned with the volatility of our (critical infrastructure) systems.

The then US Secretary of Defense, Donald Rumsfeld, expressed our inability to know our risks this way:

> ..., because as we know, there are known knowns; there are things we know we know. We also know there are known unknowns; that is to say we know there are some things we do not know. But there are also unknown unknowns the ones we don't know we don't know.
> —Transcript of DoD News Briefing, February 12, 2002

Rumsfeld was ridiculed for this, but we believe that "unknown unknowns" are a reality and that it certainly applies to complex systems.

9.7.4 Philosophical Consideration

In the article "The Singularity: A Philosophical Analysis" [53], Chalmers tries to address some of the philosophical concerns he has with the singularity. To quote from the conclusion in his paper,

> Will there be a singularity? I think that it is certainly not out of the question, and that the main obstacles are likely to be obstacles of motivation rather than obstacles of capacity. How should we negotiate the singularity? Very carefully, by building appropriate values into machines, and by building the first AI and AI+ systems in virtual worlds. How can we integrate into a post-singularity world? By gradual uploading followed by enhancement if we are still around then, and by reconstructive uploading followed by enhancement if we are not.

Chalmers' article was clearly intended to start off a debate, and it did. Many were skeptical about the prospect of a singularity, and there were also a host of other concerns and opposition.

A particularly acute question that arises from his view is how to go about "building appropriate values into machines." This is a little like Asimov's robot laws [54], which is about high-level goals. In Asimov's robot stories, there are a few that also deal extensively with the problems that arise from these laws. Ironically, it is the robots that break the laws that save humanity.

To set AIs loose in virtual worlds seems to be a better idea. The idea of using visualization and sandboxing techniques [55] is not new and it has some promise, but it is not clear that we are any good at creating properly sandboxed environments. Urgent and serious security updates for the Java VM is a frequent occurrence, and we note that a wily AI would likely only need one chance in order to break free from a virtual world. Normal accidents do happen and black swans exist.

In a follow-up paper, addressing criticism to the first paper, Chalmers continued his coverage of this topic [56]. To quote from the conclusion,

> The crucial question of whether there will be a singularity has produced many interesting thoughts, and most of the arguments for a negative answer seem to have straightforward replies. The question of negotiating the singularity has produced some rich and ingenious proposals. The issues about uploading, consciousness, and personal identity have produced some very interesting philosophy.

The "interesting philosophy" will of course equally well apply to Full HBC and Deep HBC.

9.8 Conclusions

The HBC vision is to some degree already here. Weak HBC is almost here, and it will only get better as the technologies improve.

Full HBC seems a lot more complex to achieve than mere Weak HBC. Still, brain simulation projects and similar proceed at full speed, and while it may be naive to believe that a full-fledged Full HBC interface will emerge at once, we should expect to see a gradual transition to Full HBC. To some extent, we are already into that process. For instance, we have had cochlear implant available for some time now. This implant directly stimulates the auditory nerves and completely bypasses the ear (sensor). Similar experimental devices exist for sight, but neither is providing anything near full hearing/sight yet.

Deep HBC is poised to become a reality too. This is a prediction, but if the progress in brain scanning and brain modeling continues, then it seems that it is only a matter of time before it happens. If viewing the brain as a computer is accurate, then we seem poised to find out its inner workings. Kurzweil has defined his law of accelerating return, and unless that prediction is utterly wrong, future hardware will be able to emulate the brain. It may not happen as fast as some predict, but it is of little principal difference if it takes 30, 100, or even 200 years.

Security and privacy for Weak HBC may in principle not be too difficult. To some extent it can be reduced to a technically focused engineering effort but we are coming late to the party, and security is not something that is readily bolted onto the system.

Full HBC is much more complicated. We cannot expect the current security mechanisms to apply or scale up to the challenges. The same, obviously, applies to Deep HBC.

The unknown future does indeed roll toward us, and it is high time we start to prepare ourselves for that future.

References

1 Hood, B.: *The Self Illusion: How the Social Brain Creates Identity*. New York: Oxford University Press (2012).

2 Minsky, M.: *Society of Mind*. New York: Simon and Schuster (1988).

3 Cameron, J., William, W., Jr.: Terminator 2: Judgement Day, Carolco Pictures, Pacific Western Productions, Lightstorm Entertainment, Le Studio Canal+ S.A. https://en.wikipedia.org/wiki/Terminator_2:_Judgment_Day (accessed November 11, 2016) (1991).

4 Joy, B.: Why the future doesn't need us. Wired Magazine (April). Available at: https://www.wired.com/2000/04/joy-2 (accessed November 11, 2016) (2000).

5 Pfitzmann, A., Hansen, M.: *Terminology for Talking about Privacy by Data Minimization: Anonymity, Unlinkability, Undetectability, Unobservability, Pseudonymity, and Identity Management*. Dresden: Technische Universität Dresden (2010).

6 Searle, J.R.: Minds, brains, and programs. *Behavioral and Brain Sciences* **3**(3), 417–424 (1980).

7 Turing, A.M.: Computing machinery and intelligence. *Mind* **59**, 433–460 (1950).

8 Proudfoot, D.: What Turing himself said about the imitation game. *IEEE Spectrum* **52**(7), 42–47 (2015).

9 Good, I.J.: Speculations concerning the first ultraintelligent machine. *Advances in Computers* **6**(99), 31–83 (1965).

10 Kurzweil, R.: *The Singularity Is Near: When Humans Transcend Biology*. New York: Penguin (2005).

11 Marcus, G.: Face it, your brain is a computer. *The New York Times: Sunday Review*. Available at: www.nytimes.com/2015/06/28/opinion/sunday/face-it-your-brain-is-a-computer.html (accessed October 18, 2016) (2015).

12 Marcus, G., Freeman, J.: *The Future of the Brain: Essays by the World's Leading Neuroscientists*. Princeton: Princeton University Press (2014).

13 Papakonstantinou, Y.: Created computed universe. *Communications of the ACM* **58**(6), 36–38 (2015).

14 Bostrom, N.: Are we living in a computer simulation? *The Philosophical Quarterly* **53**(211), 243–255 (2003).

15 Dick, P.K., Shusett, R., O'Bannon, D., Povill, J.: Total Recall. Movie. Available at: http://www.imdb.com/title/tt0100802/ (1990) (accessed October 18, 2016).

16 Dick, P.K.: We Can Remember It for You Wholesale. The Magazine of Fantasy & Science Fiction, pp. 4–23, April (1966).

17 The Wachowski Brothers: The Matrix. Movie. Available at: http://www.imdb.com/title/tt0133093/ (accessed October 18, 2016) (1999).

18 Kurzweil, R.: The Law of Accelerating Returns. Available at: KurzweilAI.net (accessed on October 18, 2016) (2001).

19 The Human Brain Project: Face It, Your Brain Is a Computer. Available at: www.humanbrainproject.eu (accessed October 18, 2016) (2015).

20 O'Toole, M.T. (ed.): *Miller–Keane Encyclopedia and Dictionary of Medicine, Nursing, and Allied Health*, 7 edn. Philadelphia: Saunders (2005).

21 Roche, J.P., Hansen, M.R.: On the horizon: cochlear implant technology. *Otolaryngologic Clinics of North America* **48**(6), 1097–1116 (2015).

22 Orwell, G.: *1984*. London: Secker & Warburg (1949).

23 Huxley, A.: *Brave New World, 1931*. London: Penguin (1955).

24 Postman, N.: *Amusing Ourselves to Death: Public Discourse in the Age of Show Business*. New York: Penguin Books (1985).

25 Azevedo, F.A., Carvalho, L.R., Grinberg, L.T., Farfel, J.M., Ferretti, R.E., Leite, R.E., Jacob Filho, W., Lent, R., Herculano-Houzel, S.: Equal numbers of neuronal and non-neuronal cells make the human brain an isometrically scaled-up primate brain. *Journal of Comparative Neurology* **513**(5), 532–541 (2009).

26 Kurzweil, R.: *How to Create a Mind: The Secret of Human Thought Revealed*. New York: Penguin (2012).

27 Bostrom, N.: *Superintelligence: Paths, Dangers, Strategies*. Oxford: Oxford University Press (2014).

28 Marcus, G.: *The Birth of the Mind: How a Tiny Number of Genes Creates the Complexities of Human Thought*. New York: Basic Books (2004).

29 Dehaene, S.: *Consciousness and the Brain: Deciphering How the Brain Codes Our Thoughts*. New York: The Penguin Group (2014).

30 Gibson, W.: *Neuromancer*. New York: Ace (1984).

31 Hellemans, A.: David DiVincenzo on his tenure at IBM and the future of quantum computing. *IEEE Spectrum*. Available at: http://goo.gl/oThckw (accessed November 4, 2015) (2015).

32 Sterritt, R., Hinchey, M.: Apoptosis and self-destruct: a contribution to autonomic agents? In: *Formal Approaches to Agent-Based Systems*. Lecture Notes in Computer Science. vol. **3228**, pp. 262–270. Cham: Springer International Publishing AG (2004).

33 Einstein, A., Podolsky, B., Rosen, N.: Can quantum-mechanical description of physical reality be considered complete? *Physical Review* **47**(10), 777 (1935).

34 Schrödinger, E. Discussion of probability relations between separated systems. In: *Mathematical Proceedings of the Cambridge Philosophical Society.* vol. **31**, pp. 555–563. Cambridge: Cambridge University Press (1935).

35 Køien, G.M.: Reflections on trust in devices: an informal survey of human trust in an Internet-of-Things context. *Wireless Personal Communications* **61**(3), 495–510 (2011).

36 Thompson, K.: Reflections on trusting trust. *Communications of the ACM* **27**(8), 761–763 (1984).

37 Warren, S.D., Brandeis, L.D.: The right to privacy. *Harvard Law Review*, **IV**(5), 193–220 (1890).

38 Brignall, T.: The new panopticon: the Internet viewed as a structure of social control. *Theory and Science* **3**(1), 1527–1558 (2002).

39 Køien, G.M., Oleshchuk, V.A.: Guest editorial—special edition: Privacy in telecommunications. *Telektronikk* **103**(2), 1–3 (2007).

40 Køien, G.M.: Privacy and the regulatory big brothers. *Telektronikk* **103**(2), 37–38 (2007).

41 Greenwald, G.: *No Place to Hide: Edward Snowden, the NSA and the Surveillance State.* New York: Metropolitan Books (2014).

42 Bentham, J.: *Panopticon or the Inspection House.* Lausanne: University of Lausanne (1791).

43 Køien, G.M., Oleshchuk, V.A.: *Aspects of Personal Privacy in Communications—Problems, Technology and Solutions.* River Publishers' Series in Communications, 1st edn. Aalborg: River Publishers (2013).

44 Schneier, B.: *Liars and Outliers: Enabling the Trust that Society Needs to Thrive.* Indianapolis: John Wiley & Sons (2012).

45. Carroll, L.: *Alice's Adventures in Wonderland.* Oklahoma: Brian R. Basore (1869).

46 Køien, G.M., Oleshchuk, V.A.: Personal privacy in a digital world. *Telektronikk*, **103**(2), 4–19 (2007).

47 Dolev, D., Yao, A.C.: On the security of public-key protocols. *IEEE Transactions on Information Theory* **29**(2), 198–208 (1983).

48 Vinge, V.: The coming technological singularity: how to survive in the post-human era. In: *Vision 21: Interdisciplinary Science and Engineering in the Era of Cyberspace.* vol. **1**, pp. 11–22. Westlake, OH: NASA (1993). https://ntrs.nasa.gov/archive/nasa/casi.ntrs.nasa.gov/19940022855.pdf.

49 Jenkins, W.F.: A logic named joe. *Astounding* **37**(1), 139–155 (1946).

50 Ellison, H.: *I Have No Mouth and I Must Scream.* IF: Worlds of Science Fiction. New York City: Pyramid Books (1967).

51 Perrow, C.: *Normal Accidents.* Princeton: Princeton University Press (1999).

52 Taleb, N.N.: *The Black Swan: The Impact of the Highly Improbable.* New York: Random House Publishing Group (2007).

53 Chalmers, D.: The singularity: a philosophical analysis. *Journal of Consciousness Studies* **17**(9–10), 7–65 (2010).

54 Asimov, I.: *I, Robot*. New York: Bantam Dell (1950).

55 Goldberg, I., Wagner, D., Thomas, R., Brewer, E.A.: A secure environment for untrusted helper applications: confining the wily hacker. In: *Proceedings of the Sixth Conference on USENIX Security Symposium, Focusing on Applications of Cryptography*, San Jose, CA, July 1996, USENIX Association Berkeley, Berkeley, CA, USA, vol. **6**, pp. 1 (1996).

56 Chalmers, D.: The singularity: a reply. *Journal of Consciousness Studies* **19**, 147–167 (2012).

10

The Internet of Everything and Beyond

The Interplay between Things and Humans

Helga E. Melcherts

Varias BVBA, Antwerp, Belgium

10.1   Introduction

Our society is changing at a rapid pace. For the last decades the Internet has been an intrinsic part of our daily lives. The hours we are connected to the Internet are still increasing. In 2015 almost 50% of the world population (7.2 billion) was using the Internet. Worldwide over 7 billion mobile device subscriptions circulated in 2015. In each household there are approximately 200 objects that can be connected or interconnected to the Internet. In addition the sales of wearables that are connected to the Internet has experienced a rapid growth. This shows how immense the potential of the Internet of everything (IoE) is and the possibilities linked to it. The objects and devices that are connected to the Internet will reach the number of 20 billion before 2020. Imagine all 20 billion devices generating valuable data for many industries in the private and public sectors. This data will provide information regarding new products, services, and improving supply chain and will drive companies to rethink the current business models and real-life use cases.

The 20 billion things that are forecasted to be connected do not include humans. The use of human sensors and wearables has increased substantially in the last few years. Hence the Internet of humans (IoH) is predicted to generate additional valuable data and will play an important part in the IoE and the future. Especially its relevancy for the health care and fitness industry will create interesting new business opportunities. How the IoE will further evolve beyond 2050 remains to be seen. When all things and people are connected, what is next? Human bond communication (HBC) or direct brain–machine interface (BMI)? In part 2 of this chapter, the Internet of things (IoT), IoH,

Human Bond Communication: The Holy Grail of Holistic Communication and Immersive Experience, First Edition. Edited by Sudhir Dixit and Ramjee Prasad.
© 2017 John Wiley & Sons, Inc. Published 2017 by John Wiley & Sons, Inc.

HBC, and BMI are discussed including the diverse drivers and barriers. In part 3 the emphasis is on the anticipated impact of the IoE on businesses, public sector, and consumers and describes new business opportunities. For the continuation of this paper, the assumption is made that HBC and BMI are linked to the IoE.

10.2   The Anticipated Future of the IoE

The main components of the IoE are the IoT and the IoH. In Sections 10.2.1 and 10.2.2, the IoT and the IoH are discussed in more detail. In Sections 10.2.3 and 10.2.4, the HBC concept and the BMI are explained, respectively. The drivers that trigger the anticipated development and the possible barriers are discussed in Sections 10.2.5 and 10.2.6.

10.2.1   The Internet of Things

The IoT is the interconnection of uniquely identifiable embedded computing devices within the existing Internet infrastructure. It will enable devices to make decisions, communicate, and provide humans with valuable information. No intervention of humans is needed and every action will be completely automated, also referred to as machine to machine (M2M). The IoT enables the communication with every preferred device. Every item can be connected to the Internet with a unique IP address in order to be recognized and controlled and to better analyze the generated data.

For example, our refrigerators, trash cans, washing machines, even our windows, every item you visualize, can be connected to the Internet and can generate data utilized to make our lives easier, healthier, and safer. With a simple sensor, every device can become connected. In the near future video screens will be so thin and bendable that they can be attached to every surface and to every object to provide you with information and generate data.

10.2.2   The Internet of Humans

The IoH refers to the connectivity of humans to the Internet via sensors and/ or wearable devices. Similar to the IoT, the IoH will generate an immense volume of big data through the use of wearables or mobile phones. This technology is most popular for the health care and fitness industry. The sensors/ wearables will monitor our blood pressure, blood sugar, and heart rate in real time and can be used for any diagnostics (Figure 10.1). The analysis of the retrieved data can, for example, predict a heart attack or high blood sugar. Last year GFK predicted a forecast of 51 million wearables sold worldwide in 2015.

Figure 10.1 An illustration of the possibilities for sensors/wearables.

10.2.3 Human Bond Wireless Communication

The human brain perceives the external world through its five senses and is influenced greatly by its experiences. Modern technology gave us the opportunity to communicate through technology by using only our two senses: optic and auditory. Communication with these two senses happens via camera and microphone, respectively. Being able to include the three other senses, tactile, olfactory, and gustatory, into the exchange of information is to this day not a possibility. However, Professor Ramjee Prasad from Aarhus University has developed a new revolutionary concept named HBC. HBC is an unprecedented holistic approach to describe and transmit the features of a subject in the way humans perceive it. This communication will involve all five senses for modeling the physical subject into information domain, actuating, and transmitting through communication platform. Hearing and seeing are easier to transmit since the object can be at a distance. To feel, smell, and taste, the object needs to be at arm's length for the information to be exchanged. This technology is still in its infancy but when further developed will bring transformational changes. HBC can generate groundbreaking data about our five senses, and this data can be used and set off against the IoH data, which can be extremely interesting not only for the health care practice but also for the private and public sector.

10.2.4 Brain–Machine Interface

When discussing the IoE we should also include the opportunity to directly connecting the human brain to the Internet. Will interfaces between networks and the brain be common in 2050? BMI communicating with the use of our brain should be possible in the near future according to Nigel Cameron. Can this be one of the many options for humans to communicate in the year 2050? If we assume that the BMI is going to be common, can we go as far as to envision the possibility to transfer a human's thoughts into images? Also this future technology can be of interest for numerous industries.

10.2.5 What Are the Driving Factors?

What are the driving factors for the IoE? Why is the development of communication expected to occur as envisioned in this chapter? What are the drivers triggering the IoE to happen? There are numerous drivers that can trigger the IoE to develop and these drivers vary and are reliant on the type of company, industry, or product. In this chapter the emphasis is on three drivers: technology, business requirements, and social responsibility.

10.2.5.1 Technology
- Due to the development and improvement of storage and bandwidth at increasingly lower costs, the cost of connection is reducing.
- Evolution of the Internet and the increased amounts of things and humans connected to the Internet.
- An increasing amount of devices is developed with Wi-Fi competencies and built-in sensors.
- The decreasing costs of the production of sensors.
- Development of wearables and ubiquitous connectivity.
- The fast progression of cloud and mobile computing.
- The ability to generate and to analyze big data and to utilize it in a profitable actionable way.

10.2.5.2 Business Requirements
- Increase revenue and reduce costs (improve business process logistics).
- Demand from corporations for technology to enhance productivity.

- Enhance innovation.
- Demand from corporations to improve customers experience and satisfaction.
- Request for improved marketing and advertising.
- Need to reduce the costs of health care.

10.2.5.3 Social Responsibility

- The urgency to improve health care and fitness.
- A need to improve the well-being of individuals and families.
- Human innate desire to explore unmapped terrain.
- Demand from the public for better holistic experiences.
- Request from patients to improve better health care services.

10.2.6 Potential Barriers

There are numerous reasons why the evolution to the realization of the IoE may fail. Three of the potential barriers that can delay or prevent the development from reaching full potential are technology, legislation, and financial restrictions.

10.2.6.1 Technology

The required technology such as infrastructure, wearables, sensors, and so on needs to be developed for the IoE to materialize. If the technology is not ready for the next generation, it will not reach its full potential or impact. Evolution of the underlying technology is crucial. All of the technologies in the chain need to be established in order to have effect. The IoE will not exist if the data collection is not efficient due to the failure to develop the correct smart sensors. For IoT it is fundamental to create standards and the ability to connect the different types of products with one application and to fit this technology together. The challenge is to fit every product into the IoT ecosystem.

10.2.6.2 Legislation

In order to utilize the full potential of the IoE, national and international legislation needs to be complied to. When storing sensitive personal data for health care or any reasons in a system, the data privacy laws need to be complied to. The system should enable storing sensitive data in a way that privacy and confidentiality are preserved and all ethical principles are met. Therefore, enabling data protection in the system will be an essential part of ensuring the system's compliance with the relevant national and international laws. The use of self-driving cars on the road is also subject to the development of legislation. Current European legislation does not allow a vehicle to drive on a public road without a human driver behind the steering wheel.

10.2.6.3 Financial Restrictions

Not being able to receive funding for R&D can be a significant barrier for blocking innovative technologies. Presenting a clear and precise business model and real-life use cases to present the impact and opportunity of a business idea is therefore crucial. Another barrier is the risk that the technology will not be embraced by the public. It is thus important to introduce a new product comprehensibly to the public and to properly present and inform regarding the usability of a product.

10.3   The Anticipated Impact of the IoE

The anticipated impact of the IoE will be as big as the data it will produce. It will have its impact on the private as well as the public sector and its customers. The potential for established companies and start-ups will be vast.

10.3.1   Private Sector

The IoE will enable corporations to improve revenues and cut costs. The establishment of smart factories and supply chain will improve business processes and increase productivity hence increasing revenues. Smart machineries that are interconnected not only to other machines but also to humans will create added value. Overall production costs can be reduced by, for example, using real-time inventory monitoring. Current business models will need to be reconsidered due to innovative technology and the availability of big data and information. The retrieved big data will be transferred into actionable information utilized by business leaders to make revised business decisions. Companies not able to seize the opportunities and tap into the IoE possibilities presented will stay behind.

Depending on the industry added value will be created in different shapes and forms as shown in Table 10.1. Retail will benefit greatly from the IoE by reforming its advertising and marketing strategy. Presenting a more personalized and targeted approach to advertising results in added value and increased revenues. Manufacturers will benefit greatly from advanced product lines, productivity, and supply chain improvements.

HBC, for example, can be of interest for the food and the health care industry. The ability to transfer smell and taste presents broad opportunities for the food and beverage industry.

10.3.2   Private and Public Sector

The public sector, cities, and countries will benefit by generating budget savings, superior productivity, and enhancing experiences and welfare for their citizens. In addition the IoE will create opportunities for the public sector to reduce the usage of energy and reform waste management to cut its costs. Due to the increased intelligence available, the public sector will improve its provision of information to its citizens. Smart cities will be common place and every device in the city is connected and interconnected.

Table 10.1 Industry-wide impact of the IoE.

Industry groups	2016	2050
Retailers	Advertising and marketing are often not personalized and targeted at groups of customers	Personalized advertising and marketing due to big data information
	Advertising is general and expensive	Due to the thin screens and the IoE, more attractive advertising on every surface possible. Advertisement has become less expensive and more flexible. Quickly adaptable to new market requirements in real time
	Products are general and customers have less brand loyalty. Market adaptability is slow in certain sectors	Gaining customers loyalty by offering personalized and more attractive products. Better market adaptability, more flexible to move with market trends
Manufactures	Automated machinery not always easily adjustable. Production in low-cost countries	Improved logistics and supply chain, an increase in productivity due to the use of smart machinery and applications. The use of smart applications makes the production process more flexible and easily adjustable to different business environments. Increase in revenues and lower costs
	Machinery produce expensive waste and supply chain not always efficient	Production waste is reduced to the bare minimum and machines operating time are closely planned to maximize efficiency
Automotive	Semi-self-driving vehicles allowed on the road. Communicates with driver	Smart vehicles connecting and communicating with the driver, other road users, and infrastructure. Self-driving cars
	Vehicles generate data about the vehicle. Data not always readily available to driver	Vehicles generate data about the driver and the vehicle's performance and predicts behavior of the driver. While driving it checks your health and your mood and shares information with other devices
Food	Not able to transfer taste and smell as data	HBC enables to transfer taste and smell to humans. Cooking shows on TV are able to transmit smell and taste of the food. HBC is the next step to true holistic customer/user experience
Gaming industry	Toys to life. Physical toys interact within the game utilizing near field communication (NFC). Augmented reality	Enhanced customer experience in real-life games. Human bond communication adding the possibility to actually feel the characters or objects in the game
Real estate	In homes and buildings some devices are connected	Smart homes and smart buildings. Everything inside and outside connected from the lights to the sofa or the shower controlled remotely through the cloud

10.3.2.1 Customers/Citizens

Value will be created not only for corporations but also for their customers. In order to produce better products and services for the customers, companies need to have access to additional information and to better know their customer's preferences and dislikes. Frequency of usage, information of the products' efficiency, and popularity are valuable data for companies to acquire. Due to a correct analysis of big data, companies will upgrade and develop better tailored products and services to improve customer's life and comfortability, resulting into better products and services targeted to the customer's specific needs.

Having access to sophisticated products and services will enhance customer satisfaction and customer experience, which is important for companies and for the customers. Customers and citizens will receive better targeted advertisement and information flow. Citizens will receive targeted information regarding pollution warnings, city projects, concerts, subsidies, and so on. The IoE should also result in more appropriate product charges for the customer since the production costs will decrease due to the smart factories.

HBC can be of interest for the food industry and the gaming industry since these industries tap into the holistic experience of the customer, increasing customer's experience by introducing next generation advertisements offering to smell, feel, and taste the products.

In Table 10.1 the impact of the IoE in different industries and sectors are set off against 2016.

In Table 10.2 the possible advantages and impact on the public sector in 2050 are set off against 2016.

Table 10.2 The impact of the IoE on the public sector.

Public sector	2016	2050
Countries, cities	Smart waste control	Sophisticated smart waste control. Sensor in trash can senses when it needs to be emptied. Trash cans are connected to other cans to present and predict efficient collection. Reduces costs in waste collection
	Smart screens providing information	Smart screens in the city with advertisements and providing information about, for example, pollution levels. Connecting with citizens and providing targeted information
	Semi-smart city	Smart city. Smart streetlights, smart parking indicators, energy, pollution sensors, and others are all connected and interconnected

10.3.2.2 Business Potential

The IoE will provide a boost to the economy. Many start-ups are already popping up specialized in infrastructure, smart sensors, wireless networks, cloud computing, and big data.

The IoE will produce an immense amount of data that needs to be collected transported, stored, and analyzed. The necessary technical infrastructure to facilitate this correctly needs to be developed. Correct analysis of data is vital in order to retrieve the correct actionable information for devices and humans. Opportunities for new companies to produce new smart sensors, applications, cloud, and wireless networks will increase in the coming decades.

The connectivity of all devices creates an increased security risk. Therefore consumers will demand guaranteed privacy and security of the devices and the information. Opportunities for businesses specialized in security, encryption, and privacy will increase.

10.3.3 Health Care, Patient Management, and Fitness Industry

The impact of the IoE will bring many opportunities for the health care and fitness industry specifically the IoH. For example, in health care the facility to monitor patients from a distance can create advantages for the patients as much as reducing cost for hospitals.

Generating data from patients with similar deceases and linking it to age, location, gender, and being able to store it in an international medical cloud in order to create easy access to the information worldwide will create an international database that can be accessed for all kinds of research to prevent deceases or create better medicine.

The producers of fitness products can benefit greatly from the developments in the monitoring of fitness activities. The fitness and health industry is an extensive market worldwide. Revenues are forecasted to reach $11.9 billion globally by 2020.

10.3.3.1 Anticipated Social Impact

The definition of social impact is the systematic positive effect of an activity or product on the lives and well-being of individuals and families.

The IoE will reduce the admissions and days spent in care institutions. Due to the mobility of the wearables, the health care services will be available anywhere, anytime—allowing 24/7 monitoring of patients (in particular patients with chronic diseases, elderly people, etc.), The patient can recover in the comfort of his/her own surroundings since personalized health care services are now possible anywhere and anytime the patient demands it. The patient has the opportunity to take care of herself/himself and play a more active role in the care process.

The daily activities and quality of life of older people and the chronically ill will be elevated through effective use of the IoE and better coordination of care processes. If a system can contain new easy-to-use devices with a high level of usability by elderly people, it will create added value for its users. Such a system is expected to increase the level of effective use, acceptance, and the quantity of health information. Due to the IoE the patient and the caretaker can communicate without actually having to be in personal contact.

10.3.3.2 Business Potential

The development in the health care and fitness area will create further opportunities for start-ups to develop innovative smart wearables and software ecosystems. Security requirements are essential when storing and dealing with personal data. Companies that are specialized in security (authentication and encryption) and personal data management will have great potential.

In Table 10.3 a selection of smart products for the health care, fitness, and gaming industries for the coming decades are presented.

Table 10.3 The development of smart products.

Industry	Smart products
Health and fitness industry	Wearables to monitor your sugar level, heart rate, and blood pressure and to alarm a physician if vital signs are at dangerous level
	Electronic clothes that will heat when cold and cool when warm. It can check all vital signs and calorie loss while exercising and will report to the smart fridge how much and what to eat in order to lose weight
	Smart sporting gear, sneakers, and wireless earphones for all kinds of sports are all connected and interconnected with devices and surroundings
	Advertisement based on wearables are forwarded to user. For example, if the user's fat percentage is too high, she/he will receive advertising and discounts for healthy food and gym subscriptions
Games	Virtual games to enhance fitness and health level and to improve cognitive skills when rehabilitating after surgery or an accident
Patient monitoring	Elderly or patient's wearables to be easily located and will receive information about location and connected to hospital or caretaker

10.3.4 Gender and Individuality

Big data will create possibilities for the product developer to acquire additional information about the end user. Women's immense buying force is often overlooked. Due to economic development and the increased consuming power of women consumers these days, the daily life activities of the end users are diverse and have changed. Women end users may have different preferences regarding the use of technology. It is important not to approach technology as being gender-neutral or general. In order to develop products that also appeal to women end users, we need to acknowledge these (potential) differences and research the preferences of men and women.

The IoE and the data it generates will enable the further development of products that are user friendly and will appeal to a wide(r) user base. These preferences need to be translated into more gender-specific products that can allow companies to tap into an unexploited consumer market. Engineers more than often are men for whom it can be more complex to create products for women users. However, due to the IoE the data will provide better information on the preferences of women users.

The data HBC will provide can be of interest for the food industry. Do women have different taste preferences? Is taste a factor of individuality or does gender has an effect on taste or smell? BMI when possible will take data research on gender differences between men and women to the next level. Do men and women have different thought processes? What is individuality and creativity and where does it come from in the mind? The ability to transfer thoughts into clear images will create revolutionary outcomes.

The IoE will stimulate the improvements of the gender dimension in research to become common in 2050. More women will be active in STEM offering the opportunity of more gender-balanced research group and embedded gender analysis in research. The data retrieved from the IoE will act as a driver to establish the progress in gender analysis.

In Table 10.4 a selection of industries where the gender dimensions will be present and where value can be created.

10.3.4.1 Business Potential

In the coming 2 years, the global incomes of women are expected to grow to US$ 18 trillion according to Ernst & Young (EY). In addition, women influence 75–80% of all consumer purchasing even when not paying for a product.

Table 10.4 Gender dimensions in 2050?

Industry	2050
All retailers	To enhance customer loyalty all retailers create products and advertisements that better target female consumer's preferences
All manufacturers	Female buying power has been noticed and there is more research and development done regarding female preference for objects and devices
Automotive	Design human-sensitive applications and systems in order to keep focus on the individuality of the users

Although women represent a tremendous growth market, they are still being overlooked as a separate consumer audience. Additional available research and big data on the individual preferences of the women users will enable businesses to develop more gender-prone products. Companies that are able to offer and market products for female consumers, especially in communication, have the potential to create new business opportunities.

10.4 Conclusions

The IoE can make our lives healthier, safer, and more comfortable. The anticipated impacts can be revolutionary and the new business opportunities can be numerous and profitable for companies as well as for the consumers. The predicted impacts and the opportunities, however, depend heavily on the further development of the underlying technology. Correctly analyzing the generated data to retrieve the correct intelligence is of the utmost importance. Security and privacy issues will arise for customers and for the generated data. Particularly when producing and storing sensitive personal data, authentication and encryption will be vital. Being able to feel safe and protected from cybercrime will play a vital role the coming decades. Even though the advantages are numerous, the question remains if ubiquitous connectivity is desirable. Everything from your phone to your garments will be connected. All our daily activities and humans will be traceable and visible. Taking Sections 10.2.1–10.2.4 into account, we can draw the conclusion that the technical developments of the coming decades will be interesting for the Internet and for everything else.

Further Reading

http://internetofeverything.cisco.com/vas-public-sector-infographic/ (accessed October 20, 2016).

http://www.cnbc.com/2016/02/01/an-internet-of-things-that-will-number-ten-billions.html (accessed October 20, 2016).

http://www.bbc.com/news/technology-32884867 (accessed October 20, 2016).

http://www.gfk.com/insights/press-release/gfk-forecasts-51-million-wearables-will-be-bought-globally-in-2015/; http://blog.emertxe.com/2015/11/internet-of-things-in-healthcare.html#sthash.p95YvHwT.dpuf (accessed October 20, 2016).

http://www.pewinternet.org/files/2014/05/PIP_Internet-of-things_0514142.pdf (accessed October 20, 2016).

https://blog.pivotal.io/big-data-pivotal/features/trends-the-internet-of-humans-not-things-sensors-health-fitness-healthcare (accessed October 20, 2016).

http://www.gfk.com/insights/press-release/gfk-forecasts-51-million-wearables-will-be-bought-globally-in-2015/ (accessed October 20, 2016).

http://www.huffingtonpost.com/billie-kell/the-internet-of-humanity-_b_8153438.html (accessed October 20, 2016).

https://www.theguardian.com/technology/2015/may/25/forget-internet-of-things-people (accessed October 20, 2016).

Prasad, R. (2016). Human bond communication. *Wireless Personal Communications*, **87**(3), 619–627, Springer, New York.

Pew Research Center, May 2014, "The Internet of Things Will Thrive by 2025." http://www.pewinternet.org/2014/05/14/internet-of-things/ (accessed October 20, 2016).

http://www.prnewswire.com/news-releases/fitness-equipment-market-is-expected-to-reach-119-billion-globally-by-2020---allied-market-research-504416911.html (accessed October 20, 2016).

https://blog.pivotal.io/big-data-pivotal/features/trends-the-internet-of-humans-not-things-sensors-health-fitness-healthcare (accessed October 20, 2016).

http://internetofthingsagenda.techtarget.com/tip/Internet-of-Things-IOT-Seven-enterprise-risks-to-consider (accessed October 20, 2016).

http://www.ey.com/Publication/vwLUassets/Women_the_next_emerging_market/%24FILE/WomenTheNextEmergingMarket.pdf (accessed October 20, 2016).

https://hbr.org/2009/09/the-female-economy (accessed October 20, 2016).

11

Human Bond Communications in Health: Ethical and Legal Issues

Ernestina Cianca and Maurizia De Bellis

Center for Teleinfrastructures (I-CTIF), University of Rome "Tor Vergata", Rome, Italy

11.1 Introduction

This chapter focuses on the use of human bond communication (HBC) for health applications, in particular on the ethical and legal issues that arise. For many years, the use of ICT in medicine was limited to allowing communications between remote patients and doctors (telemedicine). In the last years, there has been a rapid evolution in the use of ICT in health. The IoT framework allows a pervasive monitoring of anything around and eventually inside us and this could really open the way to novel diagnostic and therapeutic methods.

This rapid evolution has also posed several challenges as many things are not regulated yet.

What will happen when HBC will be a reality? First of all, could HBC really enable novel applications in health? And if so, would that require a novel regulation?

11.2 ICT in Health

The application of ICT to medicine starts with the so-called telemedicine, that is, medical diagnosis and management with the participants (doctors, nurses, and patients) in different places [1]. Telemedicine has come a long way from its inception in the 1960s, when NASA first put astronauts in space and "telemetered" their physiological responses from their spacecraft and space suits.

Human Bond Communication: The Holy Grail of Holistic Communication and Immersive Experience, First Edition. Edited by Sudhir Dixit and Ramjee Prasad.
© 2017 John Wiley & Sons, Inc. Published 2017 by John Wiley & Sons, Inc.

During the mid-1980s and beyond, the application of telemedicine was refined as part of disaster response and emergency preparedness. The incidences of natural disasters such as massive earthquakes and—since September 2011—the threat of nuclear and chemical hazards and bioterrorist events make it all the more important to be able to manage treatment of patients who might be remotely isolated from caregivers in traditional medical settings.

Nowadays, the use of ICT in medicine is not limited to the transmission of data but to the pervasive collection of data and also its processing and management [2]. We can distinguish between two main concepts: eHealth and mHealth. eHealth refers to the electronic storage and dissemination of health-related data and its "delicate" management from the institution side (hospital/doctors) and from the patient side who should decide who can see his/her data and what they can do with it. mHealth refers to the use of mobile and wireless communication technologies to facilitate and improve healthcare and medical services [3]. mHealth is bringing a shift to healthcare delivery. mHealth can offer significant benefits for both patients and healthcare providers, ensuring enhanced quality, efficiency, flexibility, and cost reductions in healthcare delivery. mHealth application scenarios include the active management of diseases such as diabetes, the support for independent aging to the elderly, and the monitoring of personal fitness activities to improve health and well-being. An emerging application scenario, which could offer great benefits both from the health and from the commercial point of view, is represented by dentistry applications and, more generally, intraoral sensors applications [4].

The frontier of eHealth and mHealth is the use of embedded/implanted sensors not only as sensors but also as "actuators" to automatically deliver drugs.

Both eHealth and mHealth pose many ethical and legal issues related to the use of the collected data, the protection of the data, and the liability issues when "actions" are autonomously taken.

In this paper we want to make a step forward by highlighting possible health applications of the so-called HBC concept, which is a novel concept that incorporates olfactory, gustatory, and tactile and will allow more expressive and holistic sensory information exchange through communication techniques for more human sentiment centric communication [5]. Two main applications have been identified.

11.2.1 Rehabilitation/Daily Life Support to Impaired People

HBC will definitely enable the development of more powerful virtual reality systems/robotics. Virtual reality/robotic tools are already extensively used for rehabilitation purposes and we expect rehabilitation to be one of the main applications of HBC in health. On one hand, the implementation of the HBC concept will lead to a more deep understanding on how we use our senses to understand the world around us. This knowledge is required to enable the development of systems for the exchange of olfactory, gustatory, and tactile information. Moreover, HBC will also enable the possibility to implement devices and systems that could help us to better understand these mechanisms. As an example, we are able to "perceive" the presence of an obstacle using our ears with a mechanism similar to the one for "seeing" with the eyes, that is, receiving reflected signals (or not receiving the ones that are absorbed) by the surrounding environment. However, this mechanism is little understood as we do not usually use it. On the other hand, visually impaired people usually recover this ability to sense the existence of an object that makes no sound by hearing the reflection or the insulation of other sounds in the environment, the so-called obstacle perception. A visionary research group in Japan has developed a system that has this ability of "obstacle perception." The system uses arrays of microphones. The plan is to use it for helping visually impaired people to enhance this capability faster and, hence, improve their ability to interact with the world [6].

11.2.2 Diagnosis

Most of the diagnostic tools are based on what can be "seen" (such as masses, tissue discontinuities, etc.), whatever is the type of signal or waveform that is used by the diagnostic tool. As a matter of fact, most of the diagnostic systems are "imaging systems" such as computer tomography, magnetic resonance, or ultrasound. However, it is well known that in case of "abnormal behavior/reaction of our body" or illness, our body sends also other signs that in some cases could be detected even before than a visible "object/mass." Our breath characteristics and also body's smell might change as the body also emits chemical substances that could be detected by some sensors. There could be two different approaches: (i) detecting the single substances using different sensors and (ii) recognizing the combination of these substances, that is, the smell. The latter approach requires the development of the so-called electronic nose [7]. The electronic nose is an instrument that attempts to mimic the human olfactory system. It does not identify specific chemicals within odours, but it recognizes a smell based on a response pattern. There already exist several prototypes and even few products on the market. Some studies have been already performed for diagnosis not only of cancer [8] but also other types of illness [9]. However, there is still a long way both from the sensor point of view (more powerful sensors and also for a wider range of substances) and from the data processing and classification point of view. These devices could be potentially put on mobile devices and used for personalized healthcare to provide a non-invasive means of diagnosis and monitoring.

At the same time, it would be interesting to be able to detect "diseases" or health problem by sensing substances produced by our skin or in our oral cavity, which is a kind of a "door" to the body. The ability to continuously monitor our oral cavity in a safe and noninvasive way could really enhance the current diagnostic capability, and it is an interesting and interdisciplinary research area.

11.3 Ethical and Legal Issues

eHealth technologies raise a complex set of new ethical and legal issues. Problems connected with gathering of data are not new, of course, but the rise of "big health data" and their many possible uses raise unprecedented challenges.

11.3.1 Ethical Issues

The European Group on Ethics in Science and New Technologies identified at least three key ethical implications of new health technologies.

First, eHealth, involving the electronic storage and dissemination of health-related data, affects the perception of the "self," that is, "the way the individuals view their health, their body, and conceptualize illness and disease" [10].

The same can be said about HBC. In some cases, it can lead to an "expansion" of the "self" and a revised and diminished consideration of one's illness: the case of the blind who can use a system enabling him to recognize obstacles is an expansion of one sense and also a relativization of his illness.

However, together with the potentiality, there is also a risk, since an enriched understanding of the self (both through eHealth devices and through HBC) risks becoming "detached from the social and environmental factors and from the biographical subjectivity of the patient" [10].

A second ethical implication related to eHealth is the potential transformation of the patient–physician relationship. Technological diagnosis tools, daily life support to impaired people, and rehabilitation tools can foster a significant evolution of the patient from the traditional role of a passive recipient of care. Also in this case there are potentiality and risks. As for the pros, increased technological autonomy can stimulate self-awareness. As from the cons, there are two major risks. First, these technologies could give an excessive sense of autonomy, so that medical professionals are not contacted in due terms. Second, a proper balance between autonomy and the responsibility of public authorities and the health service in general must not be lost. This is particularly evident in the cases of daily life support to impaired people and rehabilitation tools, which cannot result in a shift of responsibility to the individual and must not lead to a reduction of the standards and quality of healthcare protection.

The last point is also strongly intertwined with a third ethical implication: the one on equality and social justice. There is the risk that the use of these new technologies can make existing inequalities even more profound, since the demand would come from the well resourced and educated. Decisions by public authorities regarding investments in advanced medical technologies should not lose contact with the necessity of a proper balance with essential levels of healthcare.

11.3.2 Legal Issues

The legal challenges of both eHealth and HBC are those of privacy and protection of data and security.

11.3.2.1 Privacy and Protection of Data

eHealth relies on gathering and management of a huge number of "big health data." HBC relies on person-specific information. Both cases are challenging for the current legal frameworks and the concept of privacy.

In the EU, the right to privacy is recognized as a fundamental right in Art. 8 of the EU Charter of Fundamental Rights and Art. 16 of the Treaty on the Functioning of the European Union (TFEU), according to which personal data "must be processed fairly for specified purposes and on the basis of the consent of the person concerned or some other legitimate basis laid down by law" and "Everyone has the right of access to data which has been collected concerning him or her, and the right to have it rectified." National independent data protection authorities are competent to control compliance with EU rules for privacy.

The EU regulatory framework has long been based on the Data Protection Directive (Directive 95/46/EC of October 24, 1995) that constituted a landmark for data protection. However, there was consensus that a review of this regulatory framework was needed for two main reasons: firstly, since the legal framework was a directive (and not a regulation), it left broad space for states' autonomy in implementing EU rules, and hence national legal frameworks are very diverse; secondly, the directive was inadequate for the challenges that digitalization raises for privacy [11]. After a 4-year-long approval period, an extremely high number of amendments by EU Parliament, and the opposition of lobby groups, the new EU General Data Protection Regulation (GDPR) (Regulation (EU) 2016/679 of April 27, 2016) has been approved. The GDPR, repealing the Data Protection Directive of 1995, entered into force on May 24, 2016 and shall apply from May 25, 2018.

The GDPR changes some fundamental features of data protection regime. The most relevant innovations for purposes of eHealth and HBC are those about consent and the limits to the conservation of the data.

As a general rule, genetic data, biometric data for the purpose of uniquely identifying a natural person, data concerning health, or data concerning a natural person's sex life or sexual orientation can be processed only upon consent of the data subject. Exceptions to the consent rule include when it is necessary to protect the vital interests of the data subject or when there are reasons of substantial public interest (art. 9, GDPR). The consent principle of course applies also to eHealth and HBC.

What is changing from the past and what is of interest for regulating eHealth and HBC is the regime applicable to the processing of data for uses different from those originally considered. One of the basic principles of data protection is the one of the so-called purpose limitation, which was already part of the Data Directive of 1995. This means that data shall be processed for the purpose for which they are originally collected and that a further processing of data for purposes different from those for which consent was originally given needs a new legal basis, hence a new consent from the person concerned. However, in the big data society, the pressure to reuse data is growing, because of their commercial value [12]. The new regulation still considers the "purpose limitation" as the main principle. Nevertheless, there is an exception to this general rule: when further processing is pursued "for archiving purposes in the public interest, scientific, or historical research purposes or statistical purposes," it is considered not to be incompatible with the consent given originally (Art. 5, para. 1, let. b, GDPR). In other words, when further processing is done for scientific or statistical purposes, it is not necessary to seek again the consent of the concerned person. As far as eHealth and HBC are used for scientific purposes or statistical ones—and not commercial ones—this principle should apply also in this area.

In the case of use of data for scientific purposes, the data should normally be used through pseudonymization or anonymization (Art. 89, GDPR). However, the regulation specifies that if the scientific purposes (or other legitimate purposes mentioned in the directive) cannot be fulfilled otherwise, pseudonymization or anonimyzation can be avoided. This should also apply to eHealth and HBC: data collected through these means should be anonymized.

However, a last problematic point, as it has been previously pointed out, is that the regulation does not specify whether the reuse can be done only by the original controller of the data, or it can be transmitted to a different subject. For example, this leaves open the question whether only the physician or hospital or researcher that originally collected the data can reuse it or whether also a different hospital/research center could [12].

As for the rights of the data subject, the regulation recognizes the right of the subject to access all the personal information about him/her from the controller of the data. This shall apply to all eHealth and HBC as well (Art. 15, GDPR). This could form a legal basis for requests, which are growing, by eHealth users for easier access to their data [13].

Another innovation of the new regulation is the so-called right to be forgotten (RTBF). Taking into account the positions taken by the EU Court of Justice in the past (e.g., in the case Google Spain), the new regulation recognized to every individual the right to erasure of any information concerning him/her

(Art. 17, GDPR). No specific exception is provided for the data processed for scientific purposes (Art. 89, para. 2; GDPR does not mention Art. 17 between the provisions that can be derogated). This means that the concerned subject could explicitly withdraw its consent for the treatment of its data for eHealth or HBC.

It is important to mention that the scope of applicability of EU law has been widened with the new regulation, stating the principle of extraterritoriality. According to Art. 3, para. 2, GDPR, the regulation applies

> to the processing of personal data of data subjects who are in the Union *by a controller or processor not established in the Union*, where the processing activities are related to: (a) the offering of goods or services, irrespective of whether a payment of the data subject is required, to such data subjects in the Union; or (b) the monitoring of their behavior as far as their behaviour takes place within the Union.

Even though there are a lot of problems regarding how this provision can actually be enforced, the goal of EU institutions is clearly the one of making EU regulation applicable also to US companies, in an effort to provide an even standard of protection to EU citizens [14].

A last feature to be pointed out is that also through HBC data that could be considered as genetic can be collected. This means that international guidelines could be considered applicable. This is the case of the Council of Europe *Convention for the Protection of Human Rights and Dignity of the Human Being with regard to the Application of Biology and Medicine: Convention on Human Rights and Biomedicine* (Oviedo, April 4, 1997), setting forth a prohibition of discrimination against a person on grounds of his or her genetic heritage and the prohibition of financial gain and disposal of a part of the human body.

11.3.2.2 Security

eHealth and HBC also raise problems of security of the data, both as reliability of the data and as protection of the data.

On the one hand, there is a problem of reliability: the specificity of these data makes the distinction between trustworthy and reliable services crucial. Support from member state authorities or form EU ones, intended to certify health resources and HBC suppliers, could be advisable.

On the other hand, there are tremendous problems of security. In jurisdictions where the digitalization of data concerning health is particularly widespread, such as in the United States, there have been cases of theft of sanitary identities (more valuable than social security or credit card numbers). This makes the development of standard for securities crucial [15].

11.4 Conclusions

Health will be one of the main applications of the HBC concept, which can contribute to rehabilitation and diagnosis tools. Current use of ICT in health has already raised challenging ethical and legal issues. It must be considered that the HBC concept will even exasperate some of the ethical issues, such as the ones related to the perception of the "self." On the other hand, from the legal point of view, HBC will pose most of the legal issues related to the current use of ICT in health (privacy, protection of data, and security). The chapter provided an overview of the European legal and regulatory framework. As outlined in the chapter, despite the fact that government, regulatory bodies, and standards organizations are active in these areas, many open issues remain.

References

1 A. Amadi-Obi, P. Gilligan, N. Owens, C. O'Donnell, "Telemedicine in Pre-hospital Care: A Review of Telemedicine Applications in the Pre-hospital Environment", *International Journal of Emergency Medicine*, 2014, **7**, 29.

2 C. Chakraborty, B. Gupta, S.K. Ghosh, "A Review on Telemedicine-Based WBAN Framework for Patient Monitoring", *Telemedicine Journal and E Health*, 2013, **19**(8), 619–626.

3 E. Kartsakli, A.S. Lalos, A. Antonopoulos, S. Tennina, M. Di Renzo, L. Alonso, C. Verikoukis, A Survey on M2M Systems for mHealth: A Wireless Communications Perspective. *Sensors*, 2014, **14**, 18009–18052.

4 G. Sannino, E. Cianca, C. Hamitouche, M. Ruggieri, 2015 "M2M Communications for Intraoral Sensors: A Wireless Communications Perspective." *Future Access Enablers for Ubiquitous and Intelligent Infrastructures* (Lecture Notes of the Institute for Computer Sciences, Vol. 159), Social Informatics and Telecommunications Engineering. Cham: Springer, Paper presented at the Conference FABULOUS 2015, Ohrid, Republic of Macedonia, September 23–25, pp. 118–124.

5 R. Prasad, "Human Bond Communication", *Wireless Personal Communications, Kluwer*, 2016, **87**(3), 619–627.

6 Information Technology on Five Senses. http://www.aist.go.jp/Portals/0/ resource_images/aist_e/research_results/publications/pamphlet/today/ information_e.pdf (accessed October 20, 2016).

7 S. Chen, Y. Wang, S. Choi, "Applications and Technology of Electronic Nose for Clinical Diagnosis", *Open Journal of Applied Biosensor*, 2013, **2**, 39–50.

8 N. Kahn, O. Lavie, M. Paz, Y. Segev, H. Haick, "Dynamic Nanoparticle-Based Flexible Sensors: Diagnosis of Ovarian Carcinoma from Exhaled Breath", *Nano Letters*, 2015, **15**(10), 7023–7028.

9 M.P. Brekelmans, N. Fens, P. Brinkman, L.D. Bos, P.J. Sterk, P.P. Tak, D.M. Gerlag, "Smelling the Diagnosis: The Electronic Nose as Diagnostic Tool in Inflammatory Arthritis. A Case-Reference Study", *PLoS One*, 2016, **11**(3), e0151715.

10 European Group on Ethics in Science and New Technologies. Opinion on the Ethical Implications of New Health Technologies and Citizen Participation. https://ec.europa.eu/research/ege/pdf/opinion-29_ege_executive-summary-recommendations.pdf (accessed November 21, 2016).

11 E. Peuker, "The EU General Data Protection Regulation: Powerful Tool for Data Subjects?, VerfBlog", June 22, 2016, http://verfassungsblog.de/the-eu-general-data-protection-regulation-powerful-tool-for-data-subjects/ (accessed October 20, 2016).

12 W. Kotschy, "The Proposal for a New General Data Protection Regulation— Problems Solved?", *International Data Privacy Law*, 2014, **4**(4), 274–281.

13 See Who Owns Your Steps?, https://www.buzzfeed.com/stephaniemlee/who-owns-your-steps?utm_term=.bbeVpZw22#.tbXVO1yLL (accessed October 20, 2016).

14 C. Ryngaert, Symposium issue on extraterritoriality and EU data protection. *International Data Privacy Law*, 2015; **5**(4), 221–225.

15 Data Breaches in Healthcare Totaled Over 112 Million Records in 2015, http://www.forbes.com/sites/danmunro/2015/12/31/data-breaches-in-healthcare-total-over-112-million-records-in-2015/#5cffc4117fd5 (accessed October 20, 2016).

12

Human Bond Communication: A New and Unexplored Frontier for Intellectual Property and Information and Communication Technology Law

Edoardo Di Maggio[1] and Domenico Siciliano[2]

[1] *I-CTIF Steering Board (LAW-Intellectual Property), Rome, Italy*
[2] *Themis Law Firm, Rome, Italy*

12.1 Introduction

Human beings are able to receive and elaborate information coming from an ever-growing stream of data released by media centers, e-books, Internet, and so on. These data stimulate people's creativity and imagination and enhance sensitivity. In this way, for instance, watching the digital picture of a flower makes the person's imagination perceive its smell, as well as the picture of a chocolate bar triggers an individual's sense of taste. Wireless communication technologies, such as Wi-Fi, WiMax, and 3G and 4G protocols, have hasted the process of adaptation, which have progressively made the human being suitable to process an ever higher amount of information than it was possible to receive before, where the only way of communication was analogic or wired even if digital. Cutting wires keeping the possibility to receive data anytime in any place at high speed is giving human beings the chance to jump in a new world in which everything becomes wireless and digital. Next steps should be digital sensing and wireless human bond communications (HBC). Human perception of information has been sensed through organs as helped by a huge brain processing of that information in order to imagine and then make it as much realistic as possible. Those "conventional senses" are eyesight and hearing. Ultimately, those senses are only treated as capable of perceiving the expression of the information delivered.

In order to protect the aforementioned information, the law has progressively evolved by including protection measures that were suitable to shield the proprietary rights of both owners and users of the digital compound. As far as intellectual property (IP) law is concerned, this information is protected

Human Bond Communication: The Holy Grail of Holistic Communication and Immersive Experience, First Edition. Edited by Sudhir Dixit and Ramjee Prasad.

through copyright and trademark law, while the technologies through which the technology is delivered are protected through patent law as well. Copyright protects originality of the expression of ideas. Trademark law protects the distinctive character of a commercial sign as representative of the commercial origin of a good or service. Patent law protects inventions.

Digital and telecommunication technologies are now at a turning point, thanks to the advent of HBC: in simple words, from a legal point of view, HBC means that attorneys and legal professionals should be able to conceive in short time the framework of a smart regulation, in order to provide the principles that will be governing the interaction between human being and its environment by means of ICT technologies applied to both humans and things. The evolution that is going to occur with HBC will trigger the evolution of IP and ICT law in several aspects.

12.2 Legal Applications of HBC

As we speak, technical applications of HBC are endless and are capable of improving several aspects of the human life. Talking about improvement of life, one of the main subject areas, in which HBC will be incorporated, is health.

Health is a delicate and interesting subject for legal application because it becomes clear that regulation of certain legal aspects is fundamental in order to take the best out of any invention, progress, and, in general, any aspect pertaining to healthcare, therapeutic applications, human use, and medical devices.

HBC is related to sensitive information and works through an architecture involving (i) senducers, (ii) transducers, (iii) and, finally, human bond sensorium [1]. Human bond sensorium is the last device to be incorporated in the HBC architecture and makes it possible for the human being to experience realistic status of the matter; HBC makes it possible to incorporate sensitive technology in the body. The best explanatory example of a possible legal issue is patentability. In addition to this, another legal application of HBC is the Internet of Things (IoT), in particular, how the HBC architecture communicates, the way information delivered is protected, standards of communication, and violation of privacy.

In fact, privacy can also be involved where HBC is systematically present into medical devices. As previously mentioned, the possibility of having a medical device that delivers information on our health status is considered as a step forward for prevention of diseases. Nevertheless, privacy of this information needs to be protected. In addition to medicine, HBC applications are endless and in general involve a number of aspects that need to be taken into account.

Since the whole HBC architecture is new, new supports and technologies will be at stake in order to translate mathematical formulas into intelligible information. As Professor Ramjee Prasad mentions in his recent article, "human bond communication (HBC) understands the human sensory functionality as a system that can be expressed mathematically as the five-dimensional matrix such that, each cell of the matrix describes a physical object in terms of its sensory components [1]." This kind of "five-pronged approach" calls for five different digital supports that will lead to five or more rights. In this regard, emphasis shall be given to copyright, trademarks, and also standardization.

Moreover the "five-pronged" approach is likely to lead to five different markets, these being subject to the laws of business and economics. As it is possible to buy an mp3 song, it is also foreseeable that in a decade or less, people will be able to buy sensory wireless files or HBC files.

In addition to this, the invention of devices to be implanted in the human body for the reception and transmission of the patient's information leads to the implication of patent law.

This being said, there is a high probability that IP and related rights will be adapted in order to shadow the fast pace at which HBC technology will evolve. In particular, IP law is going to face a number of different issues, such as morality, given the overwhelming technology revolution that HBC is going to deliver to its first and central target, that is, the human being.

12.3 HBC and IoT

HBC is to be considered as a revolutionary part and evolution of the IoT concept. As a matter of fact, IoT compound possesses a broad scope. Substantially, IoT represents the way in which material devices communicate together through the Internet, a part of the machine-to-machine (M2M) communication. But be aware that in this regard, Internet involves a number of different ways of communications: in the HBC perspective the IoT shall become IoHT, Internet of Humans and Things.

In addition to this, IoT involves different devices. Among the latter it is possible to find a category that is setting the direction of this technology, namely, the so-called wearables. These devices promise to change a person's life as they track and record individual physical activity and habits and adapt to the individual needs. The latter represents the realization of Cisco chief executive John Chambers words in which IoT will "change the way people live, work, and play [2]." Needless to say that wearables are the first widespread demonstration of the capabilities of wireless technology and the IoT. The communication between a wearable device and, for instance, a smartphone is the first step forward to overtake M2M moving forward to IoHT.

As a consequence of the fast growing environment involving IoT, HBC is the next step forward to the enhancement of wireless communication capabilities. The possibility of tracking through biosensors installed inside a specific part of the human body is full of good news and promises, especially for the evolution of new medical treatments and therapies. Whether evolution is always good for technology and attracts investments, IP law can give to HBC developers a broad scope of freedom, as well as competitive advantage. This leads to the consideration of a number of issues.

12.3.1 Privacy and Data Security

HBC systems are based on similar concepts to that of the IoT. Information is delivered and translated by the architecture that involves transducers, senducers, and human bond sensorium.

Nevertheless, HBC is likely to deliver strictly personal information, such as blood pressure or the level of insulin in the organism or, in a next-step perspective, even the state of mind or the level of happiness or sadness of fear and whatever information a human being is able to process and convert to digital signals. Like this, the interaction between the various components of the HBC is regarded as a human-to-machine (H2M) architecture as it is human centric.

HBC necessitates exchange of information, which is achieved through wireless communication protocols and technologies. And since the final step of HBC is reached through at least three steps, it is possible that that information might be unlawfully accessed from the outside.

Given the huge potential of HBC in making full communication of and through human sensing possible, the unlawful access can occur at any given stage of the transmission and might involve either personal or confidential information. In this latter meaning, the information cannot only be personal but is without doubt also a secret. The Agreement on Trade-Related Aspects of Intellectual Property Rights (TRIPS Agreement) at Article 39 upholds the concept of confidential information:

> natural and legal persons shall have the possibility of preventing information lawfully within their control from being disclosed to, acquired by or used by others without their consent in a manner contrary to honest commercial practice so long as such information:

a) is secret in the sense that it is not, as a body or in the precise configuration in the sum of its components, generally in and among or already accessible to persons within the circles that normally deal with the kind of information in question;
b) has commercial value because it is secret; and
c) has been subject to reasonable steps under the circumstances, either personally, lawfully or in control of the information, to keep it secret.

It seems legitimate to expect a certain degree of protection of the information subject to HBC as letter c of TRIPS Article 39 demands. This is not necessary in order to maintain the quality of confidentiality that the information delivered has from the beginning to the end of the HBC. In general, the possibility of having technological protection measures (TPMs) serves to enhance the security and credibility of the HBC itself. Using the words of Professor William Webb, from Surrey University, on the impact of IoT on data security "we're devising a way of ensuring communications are authenticated as coming from the device as well as encrypted to avoid eavesdropping [3]." Preventing unlawful conduct that could breach security will strengthen HBC and will enable a fully integrated H2M environment. Citing again Professor Webb, "it is vital we get communications secure or else terrorists could use smart city technology to send people to a particular route or train, and pranksters could cause mayhem altering settings, such as convincing a local authority bins need emptying, when they don't [3]."

To conclude, the exchange of information occurring at different stages of HBC architecture can be an excellent area of study. It can open new scenarios of protection of that information as exchanged between two machines and lead to an ever-innovative possibility that is taking place recently in modern common law jurisdictions, namely, contracts that are formed between two machines [4].

12.4   Patents and HBC

The development of HBC will disclose a number of applications in different sectors. Given HBC's suitability to be shaped according to market and human requirements, it is foreseeable that patents will supply significant contribution to its advancement. On the other hand, patent law will trigger the generation of new and unexplored areas of interest of HBC, as based on the previous disclosed technologies.

Patents are pro-competitive and represent the way innovation is built. Just like a wall made of bricks, one after the other, patents are useful for the future, more innovative patent applications involving new inventions. The patent applicant renounces to the secrecy of its invention in order to have an exclusive right. The latter gives the patentee a competitive advantage on other market players, as he will be the only subject allowed to use that patented invention. Patents are valuable assets and allow the patentee to recover the investment he made by developing its patentable invention.

According to Article 27 of TRIPS Agreement "patents shall be available for any inventions, whether products or processes, in all fields of technology, provided that they are new, involve an inventive step and are capable of industrial application."

As mentioned in this norm, three requisites are needed in order to obtain a patent on a given product: novelty, inventive step, and industrial application. If an invention has those prerequisites, then it is patentable.

Nevertheless, there are inventions that are not patentable. The non-patent-ability follows the rules of national laws and international treaties in the way they are signed by each state. Hence, it is possible to find jurisdictions allowing a broad range of permission for patentable inventions, as well as other places with a large amount of exceptions.

The main international rule to set exceptions to patentability is Article 52.2 and Article 52.3 of the European Patent Convention (EPC). The latter states:

> "(2) The following in particular shall not be regarded as inventions within the meaning of paragraph 1:

a) discoveries, scientific theories and mathematical methods;
b) aesthetic creations;
c) schemes, rules, and methods for performing mental acts, playing games or doing business, and programs for computers;
d) presentations of information.

> (3) Paragraph 2 shall exclude the patentability of the subject-matter or activities referred to therein only to the extent to which a European patent application or European patent relates to such subject-matter or activities as such."

All of these exceptions are related to the invention "as such." This means that the principle that constitutes the subject matter of the patent application is not patentable in itself. Nevertheless, the same principle might be patentable if it is exploited together with another invention (such as a machine) in order to reach the requisite of industrial application as stated by Article 27 TRIPS Agreement.

12.4.1 HBC and Health

HBC is human centric and complies with the mission of IEEE, which is to foster technological innovation and excellence for the benefit of humanity [5]. For these two reasons, health is one of the main areas of interest of HBC. In particular, HBC will enhance the capabilities of medical devices of detecting patient's information. It is now well established that HBC's first applications in the area of medicine will be that of prevention. Sensors to be installed in the human body will gather and transmit precise information on the patient's situation. In this regard, HBC opens a new sector of telemedicine based on a brand new direct interaction between human body, communication devices, therapeutic devices, and professionals.

Beside the huge benefit for medical research and patients' health, HBC application implies patenting procedures that are enacted through patent filing within the various jurisdictions. Specifically, if the HBC device is implanted in the human body at its development stage, what implications are likely to occur for future patent applicants? The question is to be answered by looking at two aspects of patentability, namely, software patenting and morality.

12.4.2 Software Patenting

It shares with IoT interchangeability of information, which occurs through waves, transmitted and translated between the three HBC stages. The optimization of this process is carried out by a software. It is likely that software will play a significant role in the process of HBC delivered information. It will need to process a huge amount of information in terms of thousands of terabytes.

Software patenting is a controversial issue, and a medical device might perform tasks that are already in the state of the art, given the higher level of competition in the pharmaceutical and medical sector. Nevertheless, the so-called computer-implemented inventions are often patentable. It is appropriate to take in consideration software patenting under the light of two different jurisdictions, which represent the inspiring sources of patent law: European Union and the United States.

It is a matter of fact that EPC excludes computer programs from patentable inventions. It does so in Article 52.2 letter c, where it states that "schemes, rules, and methods for performing mental acts, playing games or doing business, and programs for computers" cannot be patentable. In this regard one may be confused as to the outcome of this paragraph. In reality that exclusion represents an *incomplete truth*. And Article 52.3 EPC completes the circle. In fact by stating that "paragraph 2 shall exclude the patentability of the subject-matter or activities referred to therein only to the extent to which a European patent application or European patent relates to such subject-matter or activities as such," it means that the single object or source codes of a computer program are excluded from patentability. Codes are in fact algorithms and, as such, mathematical formulas and hence not patentable.

But if the computer program is the part of an invention or simply a device involving an inventive step, together with that computer program, it is then possible to reach patentability. European jurisprudence teaches exactly this lesson. As a confirmation of the latter statement, in the case involving the software house Vicom, the European Patent Office (EPO) Board of Appeal states: "paragraph 2[1] shall exclude the patentability of the subject-matter or activities referred to therein only to the extent to which a European patent application or European patent relates to such subject-matter or activities as such."[2] This concept is further strengthened by the UK Board of Appeal that calls for a contribution to the known art in order for the invention to be patentable [6]. In what

---

1 Paragraph 2 of Article 52 EPC.
2 VICOM/Computer-Related Invention T208/84 [1987] EPOR 74.

does consist this contribution to the known art? In particular, what does a computer program need to satisfy in order to be patentable?

The answer to those questions resides in various formulas elaborated by the courts. One of those formulas can be concretized in the "technical effect approach," elaborated by the EPO Technical Board of Appeal in the case involving IBM.[3] In light of the technical effect approach "...a patent may be granted not only in the case of an invention where a piece of software manages, by means of a computer, an industrial process or the working of a piece of machinery, but in every case where a program for a computer is the only means...of obtaining a technical effect within the meaning specified above, where, for instance, a technical effect of that kind is achieved by the internal functioning of a computer itself under the influence of the said program. In other words, on condition that they are able to produce a technical effect in the above sense, all computer programs must be considered as inventions within the meaning of Article 52(1) EPC, and may be the subject-matter of a patent if the other requirements provided for by the EPC are satisfied." According to Vicom and IBM decisions and other jurisprudence,[4] the EPO has developed its own guidelines for the patentability of computer programs, calling for the indispensable presence of the inventive step in order to progress in the valid examination of the patent claim.[5]

As regards the United States, the situation of software patenting is different and more freely interpretable. In fact, firstly the United States does not provide exceptions to patentability. Secondly, the US case Diamond v. Chakrabarty represents the corner stone of US patentable subject matters. In this decision, the supreme court affirmed that everything made by man under the sun is patentable. Hence there is no doubt any invention is virtually patentable, even if immoral or against ordre public.

For computer programs the case *In re Bilski* seems appropriate to describe US software patenting. The case introduces the machine or transformation test. This test can be summarized as follows. In order to be patentable, a

---

3 EPO Technical Board of Appeal in IBM/Computer Programs T935/97 [1999] EPOR 301.
4 Hitachi T258/03 EPO Technical Board of Appeal.
5 "If claimed subject-matter does not have a prima facie technical character, it should be rejected under Article 52(2) and (3). If the subject-matter passes this prima facie test for technicality, the examiner should then proceed to the questions of novelty and inventive step. In assessing whether there is an inventive step, the examiner must establish an objective technical problem which has been overcome. The solution of that problem constitutes the invention's technical contribution to the art. The presence of such a technical contribution establishes that the claimed subject-matter has a technical character and therefore is indeed an invention within the meaning of Article 52(1). If no such objective technical problem is found, the claimed subject-matter does not satisfy at least the requirement for an inventive step because there can be no technical contribution to the art, and the claim is to be rejected on this ground." EPO Guidelines for Examination C-IV: Programs for computers.

computer program must be tied to a particular machine or apparatus (i) and must transform a particular article in a different state or thing (ii). The two phases are inextricable as the computer program must be a part of a physical machine that transforms something into something else.[6]

With HBC it is likely to assist to such a transformation, at least on the side of the information, which is transformed into sensitive and realistic data. Patent implications on HBC will be likely to incur where the computer program is used in a device that senses diseases and transforms chemical compounds into sensitive and intelligible information. The information in itself will not be patentable, and it is more likely that will be subject to confidential measures in order to protect the privacy of the patient. Nevertheless, the interaction between the HBC medical device and software is likely to make the two components, considered as a unique patentable invention, given it satisfies the three criteria of patentability of Article 27 TRIPS Agreement.

12.4.3 HBC for Health and Morality Issues

Is HBC applied to human body immoral? This is a question that, for the large part, regards EU patent law, as the United States lacks exclusions of patentable subject matters. In the EU as well as in other member states of the EPC, if the invention is immoral or contrary to the *ordre public*, it is not patentable.[7] There are a number of issues related to the implantation, whether invasive or not, of a medical device in the human body.

In principle, it must be said that HBC is something that has never been seen before within the patent landscape. So, it is possible that new laws will be drafted in order to better fit HBC with the concept of patentable subject matter. This being said, morality is a broad concept, and it is best defined by the EPO Board of Appeal that states "the concept of morality is related to the belief that some behavior is right and acceptable whereas other behavior is wrong, this belief being founded on the totality of the accepted norms which are deeply rooted in a particular culture. For the purposes of the EPC the culture in question is the culture inherent in European society and civilization. Accordingly, under Article 53(a) EPC, inventions the exploitation of which is not in

---

6 In re Bilski 545F. 3rd 943 (Fed. Cir. 2008).
7 The EPO guidelines to patentability follow Article 53(a) EPC: "Any invention the commercial exploitation of which would be contrary to "ordre public" or morality is specifically excluded from patentability. The purpose of this is to deny protection to inventions likely to induce riot or public disorder, or to lead to criminal or other generally offensive behaviour...Antipersonnel mines are an obvious example. This provision is likely to be invoked only in rare and extreme cases. A fair test to apply is to consider whether it is probable that the public in general would regard the invention as so abhorrent that the grant of patent rights would be inconceivable. If it is clear that this is the case, objection should be raised under Article 53(a); otherwise not..."

conformity with the conventionally accepted standards of conduct pertaining to this culture are to be excluded from patentability as being contrary to morality."[8]

In order to demonstrate that HBC applied to human being is in itself desirable and then moral, it is sufficient to exclude the latter from the categories of non-patentable inventions that could represent a threat to health and wellness. At the same time we might ask ourselves if "European society and civilization" will benefit from a device that actually increases the possibility to spot a given disease in an early stage of its formation or that is capable to implement a given therapy.

The first norm that needs to be taken into account is Article 5 of EU biotechnology directive.[9] The first paragraph of this norm impedes the patentability of "the human body, at the various stages of its formation and development, and the simple discovery of one of its elements." In this sense, it must be noted that the HBC device is implanted surgically into the human body when the patient is in all its faculties. Clearly, the HBC device is not an isolated part of the human body as, at the moment, it is not considered as biologic material. In this last example it would have surely be patentable according to Article 5.2 of the EU biotechnology directive. But according to Article 5, nothing stands against the patentability of the HBC devices. They are not an integral part of someone's body and represent therapeutic and monitoring machines.

Another norm that is worthy of attention is Article 6 of EU Biotechnology Directive. In particular, this norm sets the balance between animal or human suffering and medical benefit. The first paragraph of Article 6 deals with the concepts of non-patentability of inventions that are contrary to morality (or *ordre public*). On the other hand, the second paragraph of Article 6 specifies what is included in paragraph 1. As points (a), (b), (c), and (d) of the second paragraph imply radical transformations of the human and animal body, it is unlikely that HBC devices may encounter any obstacle to patentability.[10] It is instead probable that HBC-implemented inventions will be considered as other medical devices are considered. After all, if pacemaker was patentable, HBC devices will be as well.

---

8 EPO Technical Board of Appeal in Plant Genetic Systems/Glutamine synthetase inhibitors T 356/93 [1995] EPOR 357.

9 Article 5 EU Biotechnology Directive 98/44/EC of July 6, 1998.

10 Article 6.2 EU Biotechnology Directive 98/44/EC of July 6, 1998 states "2. On the basis of paragraph 1, the following, in particular, shall be considered unpatentable: (a) processes for cloning human beings; (b) processes for modifying the germ line genetic identity of human beings; (c) uses of human embryos for industrial or commercial purposes; and (d) processes for modifying the genetic identity of animals which are likely to cause them suffering without any substantial medical benefit to man or animal, and also animals resulting from such processes."

Also, it must be remembered that surgical methods are not patentable according to Article 53(c) EPC.[11] Nevertheless, the case G 1/07 of the EPO adds also another aspect of patentability related to medical devices. The EPO considered as patentable "methods concerning only the operation of a device, without any functional link to the effects produced by the device on the body...." This is good news for HBC. As a consequence it is possible to obtain a patent on the method of functioning of a device, provided that that method does not extend to any actual link capable of causing direct effects of the human body. And this concept is valid both where a therapeutic treatment is involved and in cases where the implantation of the device requires a surgical step.[12] This is the last and authentic demonstration that HBC medical devices are indeed patentable, whether implemented by a computer program or alone.

12.4.4 HBC and Standards

HBC involves interoperability. An important issue related to HBC is the communication technology that is likely to be used in order for HBC devices to communicate. In this regard, standardization is highly desirable. Standard essential patents (SEPs) are necessary rights that need to be breached in order to allow a given product to be developed.

In particular, since HBC devices are going to be patented, if they have proprietary communication technology, other devices from different manufacturers will be hampered in their functions.

During the first period of introduction of HBC technology, interoperability between devices might lead to anticompetitive behavior. It is possible that while dedicated standards are not approved or inexistent, some companies will take advantage of this uncertain situation [7].

The means of competitive advantage are represented by licensing contracts, which will be indispensable for third parties in order for their HBC devices to join HBC architecture. High licensing rates are likely to occur during the first round of marketing of technology. As a consequence the high price of technology licenses will represent an obstacle to development and a betrayal of the spirit of patent law itself.

For this reason, ETSI and other organizations such as oneM2M.org promote the creation of essential standards and SEPs [8]. By recognizing the importance and contribution of patent owners, those standard-setting organizations will set pro-competitive fair, reasonable, and nondiscriminatory (FRAND) rates.

---

11 "Methods for treatment of the human or animal body by surgery or therapy and diagnostic methods practised on the human or animal body..." shall not be patentable.
12 "G 1/07 and the exclusion from patentability of surgical methods at the European Patent Office" [2010] J Kemp & Co.

12.5 Conclusions

In conclusion HBC technology is full of promises, as it will develop in a number of different and innovative applications.

HBC is a valid instrument for the improvement of quality of life despite law, and regulations should be developed in order to include any possible HBC application. In order to reach the ambitious goals delivered by HBC, information and communication technology, IP, and the law will need to develop coherently. Hence, a cross-disciplinary approach is the key factor for reaching full implementation of the legal and general content of HBC. A concept that is moral, lawful, and able to improve quality of life and the relationship between human being and its surrounding environment.

In order for these principles to be implemented, a solution to FRAND licensing rates and standardization is the main factor to be taken into account. If harmonization of the law, fair use of patented inventions, and other IP rights, as well as standards, is found, HBC will gain a prominent and indispensable dimension starting from the next decade.

References

1 Ramjee P, "Human bond communication," *Wireless Personal Communications*, vol. **87**, no. 3, pp. 619–627, 2016.
2 England P, "Patent Issues and the Internet of Things" (2016). Available at: http://united-kingdom.taylorwessing.com/download/article_patent_iot.html (accessed June 27, 2016).
3 Hargrave S, "How Will the Internet of Things Impact Data Security?" (*The Guardian*, 2015). Available at: http://www.theguardian.com/small-business-network/2015/jul/30/internet-of-things-data-security (accessed June 27, 2016).
4 Nolan P and Adair M, "The "Internet of Things": Legal Challenges in an Ultra-Connected World" (2016). Available at: http://www.mhc.ie/latest/blog/the-internet-of-things-legal-challenges-in-an-ultra-connected-world (accessed June 27, 2016).
5 IEEE Strategic Plan 2015–2020 (2015). Available at: https://www.ieee.org/about/ieee_strategic_plan_2015_to_2020.pdf (accessed June 27, 2016).
6 Fox LJ. *Merrill Lynch Inc.'s Application*, RPC 561 at 569 (1989).
7 Diaz Alaminos B and Oker-Blom M, "The Internet of Things and Intellectual Property Rights" (2016), *The Trademark Lawyer*.
8 Dahmen-Lhuissier S, "ETSI—How We Work, Organization, Rules, Charts, STFs, editHelp!, IPRs, Effective Participation, Start New Activities" (2016). Available at: http://www.etsi.org/about/how-we-work (accessed June 28, 2016).

13

Predicting the Future of ICT: A Historical Perspective

Silvano Pupolin

Department of Information Engineering, University of Padua, Padua, Italy

13.1   Introduction

In this chapter several aspects related to wireless communications are analyzed. The success of wireless is mainly due to two causes:

1) The change of communication paradigm from station to station to person to person
2) The technological advancement that transformed the telephone terminal in a multimedia terminal processor

Both elements are revolutionary and give answer to the needs of people. This successful story was incredibly fast (three decades).

What is next for wireless communications?

Current terminals are able to give all we need from communications. Next solutions have to find new needs of the persons. To have a clear view of these new needs, we should consider integrating research teams composed of researchers with completely different competences to find new and interesting applications.

Many new research fields are being sought and they could be split in two families and shortly summarized as follows:

1) Technological research:
    a) Hands-free terminals
    b) Augmented reality
    c) Green terminals and systems

2) Application research:
    a) Remote support for disabled people with guidance
    b) Remote support for elder and diseased people

Human Bond Communication: The Holy Grail of Holistic Communication and Immersive Experience, First Edition. Edited by Sudhir Dixit and Ramjee Prasad.

c) Mobility support
d) Road security
e) Mobile office

Most likely success of both research fields will mainly depend on technical research on hands-free terminal (1a). Secondly, other features, for example, fashion design, will create new client needs.

Next decades' research developments on ICT will be related to interaction with multidisciplinary teams, which look for new interesting and useful applications. One of the most challenging activities will be related to health and human support where ICT will bring new engineering solutions. Another application will be related to self-driving cars in order to guarantee a safe person transportation to his/her final destination even if the person does not have a driving license.

13.2   A Short Run-Through Technology Evolution

World War II (WWII) was a push forward to the development of new telecommunication services, mainly for military applications. The evolution of radar allowed Western Alliance to identify enemy ships as well as airborne and to take appropriate offensive and defensive measures. Radio broadcasting was used to inform population about current situation and the partisans on action commands.

At the end of the war, the impressive technological progresses were employed to bring new improvements and services during reconstruction of European cities that were almost completely destroyed.

ICT developments were particularly impressive within a decade from WWII.

Seven basic technologies acted together as a positive feedback to push the advance of ICT forward: (i) microelectronics, (ii) digital signal processing, (iii) computer engineering, (iv) computer networks, (v) telecommunication systems, (vi) telecommunication networks, and (vii) radio communications.

Here we look to the past progresses that brought the solutions close to theoretical performance limits. Further advances require new research studies in different disciplines. The next paragraphs will be devoted to the actual evolution where disciplines merge together to continue their development and others have finished their run and disappeared. New disciplines rise to support either the requirements of clients or completely new applications.

13.2.1   Microelectronics

The fundamental discovery in microelectronics is represented by the transistor invented in 1950 by W. Shockley, J.S. Bardeen, and W.H. Brattain [1–3] at Bell Telephone Labs, which improved a lot the power consumption and the reliability with respect to vacuum tube amplifiers. Since then, solid-state electronics moved forward astonishingly starting from integrated circuits (IC), independently discovered by J. Kilby at Texas Instruments (TI) and Robert Noyce at Fairchild Semiconductor [4, 5]. Afterward technology moved to large-scale integration (LSI), very-large-scale integration (VLSI), and so on up to the current system on chip (SoC). This technology triggered the development of modern systems.

13.2.2 Digital Signal Processing

After WWII the basis for the signal digitalization has been posed with the key paper by Oliver, Pierce, and Shannon in 1948 [6]. Since then, the advance of the digital world was impressive even using the limited computational power made possible by microelectronics. Indeed, a decade later the paper by Oliver, Pierce, and Shannon led to the implementation of the PCM in the telephone system [6]. Two decades later (1980) Philips and Sony developed independently the compact disc, which revolutionized the way to reproduce sounds [7, 8]. After another decade the available computational power enabled techniques for real-time redundancy reduction and made possible the digitalization of video signals. The first successful experiment was the satellite broadcasting by digital high definition TV of the 1990 world soccer championship matches. The further development of this experiment was the definition of the worldwide standard MPEG-2 for the digital television [9]. Then, MPEG standard further improved including the transmission of video images through Internet at several different rates and image qualities.

The development of digital signal processors (DSPs) pushed forward the signal processing field, providing high computational power for general-purpose hardware at low cost.

13.2.3 Computer Engineering

Computer engineering took advantage of the progress of microelectronics and the increase of speed and complexity of digital electronic circuits. The design of microprocessor in the early 1970s became a relevant turning point for the spread of computers. Ray Holt and Steve Geller designed the MP944 chipset in 1970 as the main flight control computer in the new F-14 Tomcat fighter (R.M. Holt, Architecture of a microprocessor, unpublished work 1971, https://docs. google.com/viewer?url=http://firstmicroprocessor.com/documents/ap1-26-97.pdf) whose work was classified by the US Navy. In 1971 Gary Boone designed for TI a microprocessor named TMS1000 [10], which went into the market in 1974, and in the same year Federico Faggin designed the Intel 4004 [11], the first four-bit microprocessor commercially available starting 1972. Within few years computational power increased and the number of bits were brought up to 64. This promoted the development of personal computers

(PCs), tablets, smartphones, and so on. TI and Intel specialized their microprocessor products toward two distinct applications: digital signal processing and general-purpose microprocessors, respectively. Since then microprocessors became pervasive devices, which today can be found at least in all commercial electronic products.

The effect of microprocessor was to introduce PC in offices and homes. In offices the change from mainframe to PC forced the need of a data communication network to connect all the PCs in order to share documents generated by different PCs. The computer industries standardized the series IEEE 802.xx data communication protocols to connect computers within a society premises in an efficient and reliable way. These protocols considered both wired and wireless links. Such communication system was named local area network (LAN). LANs located in different buildings were interconnected through dedicated leased public data communication links.

13.2.4 Computer Networks

Based on a concept first published in 1967, ARPANET was developed under the direction of the US Advanced Research Projects Agency (ARPA). In 1969, the first network interconnected four university computers. The initial purpose was to communicate with and share computer resources mainly among researchers at the connected institutions. ARPANET took advantage of the new idea of sharing a common transmission medium among several users by sending information split in small units called *packets*. Each packet had a header containing the sender and destination addresses so that at each network node, the packet was routed toward its final destination. At destination packets were merged together to reconstruct the original message. The development of the TCP/IP protocol in the 1970s made the expansion of the size of the network possible, which nowadays had become a network of networks, in an orderly way.

In the late 1980s CERN, the European Organization for Nuclear Research, had severe problems to share information among different computers because data were stored using different formats and computer programs. By October 1990 Tim Berners-Lee, CERN software engineer, wrote the foundation of today's web by using three fundamental technologies:

- HTML: HyperText Markup Language, which is the core language to create documents on the web.
- URI: Uniform Resource Identifier. The definition of URI from the IETF URI WG is "Uniform Resource Locators (URLs) for encoding location and access information across multiple information systems, Unique Resource Serial Numbers (URSNs) which specify a standardized method for encoding unique resource identification information for Internet resources, and Uniform Resource Identifiers (URIs) which specify a standardized method

for encoding combined resource identification and location information systems to be used for resource discovery and access systems in an Internet environment."

- HTTP: Hypertext Transfer Protocol, which allowed for the retrieval of linked resources from across the web.

By the end of 1990, the first web page was served on the open Internet, and in 1991 people outside CERN were invited to join that new web community.

Immediately the web community produced the following revolutionary ideas to enter the web:

- *Decentralization*: No permission was needed from a central authority to post anything on the web; there was no central controlling node and thus no single point of failure and no "kill switch!" This also implied freedom from indiscriminate censorship and surveillance.
- *Nondiscrimination*: Different quality of service (QoS) networks could interface with each other, despite their difference in quality and cost. This principle was also known as *net neutrality*.
- *Bottom-up design*: Participation of everyone was encouraged so that not only small groups of experts wrote code for web activities.
- *Universality*: Multilanguage protocols and dictionaries were developed to allow everyone to publish everything on the web, despite diversities of language, culture, or political beliefs.

13.2.5 Telecommunication System

Since its beginning telecommunication systems were split in two different networks: telephone and telegraph. The first was designed for voice transmission, while the second was devoted to the transmission of text (basically was a digital transmission). Telephony moved from local communications to a worldwide interconnection. An astonishingly technological improvement led to the development of the analog telephony as it will be shown in the milestones as follows:

March 1876—Graham Bell patented his telephone handset.
December 1877—The first telephone company used electric telephones.
January 1878—First manual telephone exchange in New Haven (NJ).
February 1878—Telephone call between Menlo Park (NJ) and Philadelphia, 210 km away.
March 1891—Strowger patented the first automatic telephone exchange.
January 1907—De Forest patented the triode vacuum tube.
April 1911—Multiplex FDM [12].
1915—Crossbar switch.
June 1918—Commercial multiplex FDM in service [13].
April 1924—Minimum bandwidth required for telegraph transmission [14].
January 1932—Amplifier stabilization [15].

October 1938—H. Reeves patented the PCM in France [16].
February 1942—H. Reeves patented the PCM in the United States [16].
July 1948—Shannon published the fundamentals of communication theory [17].
November 1948—PCM theory presented [6].

Theoretical and practical performance of PCM was developed before WWII. The system however appeared too complex for the available technology, and its reliability was too low to be actually used. The invention of transistor in 1949 changed several aspects, and a decade later in 1963, the first PCM line was installed [18]. This was the first step of the digital revolution.

PCM at the beginning was mainly a transmission system but quickly moved to consider other aspects as signal multiplexing and signal processing. However, technology was the main driver for the design of new services. The first digital application was interactive video communications. Due to technology limitation, such as a very limited bandwidth and processing capability, the quality of the service was extremely poor. The consequence was that the service dropped out in a short time.

13.2.6 Telecommunication Network

Because of the high cost of transmission infrastructure, telecommunication network was designed for telephony and telegraphy, separately. In order to reduce costs, the network was designed with a hierarchical structure. Since 1918 for long-distance telegraph transmissions, 24 telegraph lines were multiplexed on a unique signal that was transmitted over a single pair of copper wires. The signals were separated in frequency by using different carriers, the so-called frequency division multiplexing (FDM). A similar system was used a decade later for the telephone signal multiplexing. The number of multiplexed telephone channels increased with time up to 10,800 channels (last release was in 1980). Telegraphy, a low bit rate digital transmission, evolved to use the telephone channels as support for transmission in order to reduce costs.

Telecommunication networks were designed to support voice and, in order to reduce global costs, were organized hierarchically. Network nodes were the switching centers that were classified as class switches with numbers going from 1 to 5, being Class 1 the national and international tandem gateway, Class 2 the regional, Class 3 the primary, and Class 4 the toll center tandems. Class 5 is the end office where subscribers were connected. In this way each tandem switch demultiplexed signals arriving from other switches and commuted toward their next destination by multiplexing them in order to reduce the number of lines needed. Technology changed the way in which switch was done but the philosophy remained the same: *circuit switching.* Incidentally, Internet will then have further changed to *packet switching.* During the past decade the service integration and the request of multimedia services forced telecom manufacturers and operators to move toward the use of Internet because it was the only protocol defined to support multimedia communication. However, this choice required many adjustments in the use of the Internet because telephone services were of excellent quality, while Internet did not guarantee QoS.

13.2.7 Radio Communication

Radio communications were fascinating since the first Marconi's wireless telegraphy system commercially available in 1894. Since then, radio communications evolved to radio and television broadcasting and to radio communication services. Big improvements in the radio systems were obtained with the use of vacuum tube amplifiers (1920), which enabled the design of new radio receivers. Later, after WWII, the use of transistors led to the manufacturing of portable receivers. Further improvements in the radio communications arose from using microelectronics, signal processing, and computer computational power gains in the last three decades. Nowadays common sense about radio communications is related to mobile communications, that is, systems that allow us to use data and voice services always and everywhere. Young people believe this telecommunications infrastructure and service has always existed, but it was actually designed and placed first in service only in 1973 [19–21]. This new technology changed the communication paradigm from a *station-to-station* to a *person-to-person* mode.

Next section deals with progress in the last decades. Later, the evolution perspectives of wireless communications will be considered.

13.3 Telecommunication Evolution in the Digital Era

In the last three decades telecommunication systems moved from separated networks dedicated to a well-specific media (telephony network, telegraph network, high speed data network), where each network took care of the required QoS of the specific service, to one multimedia network where several different services with different requirements shared the same communication network. This change was supported by the technology advancement and today we are seeing the effect of this revolution.

In 1980s when digitalization of the transmission network was almost completed, the idea of an Integrated Service Digital Network (ISDN) was born. This idea was not complete because data network was at its dawn and service integration was referred to the telephone and telegraph services plus some new services, as video (not yet available). The idea was good but the design shortly showed to be not adequate. Indeed the final user capacity was limited to 160 kbit/s and the system was not scalable. Few years later it appeared insufficient to support the planned services to the final user.

It is worth noticing that in the same period, Internet were booming moving to the World Wide Web (WWW) as well as the new mobile communication network, which was designed at the beginning only for voice (completely separated from the ISDN). In the same period fiber optic communications began its service and a continuous increase in data rates analogous to the increases of the number of transistors in an IC (Moore's law) started. Specifically, link capacity doubled every 2 years (Keck's law) so that in 35 years 100 Tbit/s was on a single fiber.

Service integration was realized together with the network integration. Indeed during the last decade of the 20th century, an all IP long-distance telephone network was designed in order to take care of the QoS. Furthermore, mobile and fixed telecommunication networks converged, and mobile backbone network became also the fixed network of the operator. This convergence was completed with the standardization of the 4G mobile network, which was designed as all IP network. Thus, it supported multimedia messages with QoS. With an all IP network new terminals and services appeared. In the last decade tablets and smartphones allowed a single user to be always connected everywhere in the world. New services appeared as applications (APPs) that could be downloaded into the terminals, and a booming of data services had started. It is interesting to remark that technology nowadays is no longer pushing the innovation but applications and people needs are driving advancements. These facts are changing completely the perspectives related to the new research and development of telecommunication systems and services during the 21st century. Moreover, business model had been changed because it is no longer managed by developing new technologies by the telecom operators but by new services proposed and managed by new societies. The effect is that telecommunication network and technology, in general, had become a commodity, and revenues for the network operators are related only to the transport of information and not to new services carried on anymore.

13.3.1 Wireless Technological Development

The increase in data communication in the last decade required a new network design to support the customer requirements. The customization of the terminal is the last revolution due to mobile networks and the presence of thousands of APPs that are tailored to specific needs of the clients.

Old telecom operators (OTOs) used to deploy new technologies and are experiencing a rapid change of their business. The new APP service business requires new competencies in order to maintain the revenues. Recently, many of the OTOs signed agreements with new societies designing APPs, providing them network capacity and QoS in order to make APP work properly. In the recent past OTOs created new societies to manage and design the network while they more concentrated on services. On this way in few years, one society can be expected to manage the whole telecom infrastructure, while the OTOs could offer services to the travelling clients, on a common infrastructure, on the basis of their needs (real or induced).

In the future two main areas are expecting to largely expand: (i) technology novelties and (ii) new applications.

As technology regards, hands-free terminals, augmented reality, and green terminals and systems are three critical aspects of the near future. The main characteristics related to each one will be shortly presented in the following.

13.3.1.1 Hands-Free Terminal

The progress of voice commands targets an intelligent terminal without any display. In this way terminal could be very small and could use human bones to transmit and receive voice signals. Signals would be less corrupted by ambient noise, more clearly received, and less power would be needed so that battery would last longer. However, such a terminal is not yet able to cover the whole set of today's services offered by a smartphone. For instance, we use an agenda to search for a phone number and we look through it to find names and phone numbers: a display showing a standard lookup table is still necessary. We need a new way to display data and scroll them. The same requirements are for games where we use fingers to hit the keyboard or the wireless controller. New interfaces and new human attitudes to access services are required. The future could be to capture human movements and interpret them as commands. This will require an intelligent system able to learn a person's gestures and translate them into appropriate commands for the terminal.

13.3.1.2 Augmented Reality

The display is useful for several applications and it could be associated to glasses with augmented reality. We could display on them what we typically see on the terminal display and some more. In this way we could help clients to perform operations by showing in real time what he/she has to do. Elders have problems in identifying where they are, and a clear support where to go will help them a lot. With augmented reality it is possible to display where to go. Further applications could be developed to support the requirements of clients when moving or relaxing. For the next decades we will see standard Internet video applications to move to augmented reality where, following the example illustrated earlier, the route we have to follow is overlapped to what we see. This could be done in two different ways: first, by using eyeglasses where the road to follow is projected onto and, second, by capturing the street view with two video cameras, mixing them with the road directions and showing this composite signal on a screen embedded in eyeglass case. Many other unexpectable solutions will appear by 2050.

13.3.1.3 Green Terminals and Systems

Until today we saw a dramatic run to increase telecommunication link capacity at a cost of increasing the required energy/bit. The result is an increasing energy requirement to support telecommunication systems. Now, the technology is looking at low power devices able to operate by harvesting the needed energy from several different renewable sources, for example, solar, wind, mechanical, and thermal. The design of these new terminals is based on a new paradigm: minimize the required energy/bit to transport the information at destination. New theoretical studies as well as the development of new hardware are expected. These new terminals will be the basic block for the Internet of things where a very large number of sensors will be deployed and interconnected with a very low probability either to be connected to a power source or to change batteries once discharged. The final result is a reduction of power consumption by telecommunication systems. This is an interesting research area for 2050 and beyond.

13.4 Telecommunication Evolution: From Technology to Applications

In the last decades technology progress supported the evolution of telecommunication paradigms from a station-to-station to a person-to-person communication and from voice to multimedia. At the same time networks evolved from specialized networks for a specific service to a unique multimedia network. Recently, telecommunication evolution appears to be more and more centered on the needs of people. This phenomenon was driven by the definition of worldwide standards and the production of millions of terminals whose cost collapsed, and many new clients are now able to buy it and change it every few years. Powerful multimedia terminals are now available on the market, and they are commonly used to browse Internet to search for something, for example, a restaurant or a museum; to navigate a map; or to sense biological measures, for example, heartbeat rate, blood pressure, and body temperature.

A new market appeared recently related to the design of APPs devoted to specific applications for the terminals. These APPs are related to obtain a service, e.g. to get a taxi cab, check if a parking lot has free slots, etc.

Therefore, future wireless service development will be based on human needs, as:

1) Remote support for elder and diseased persons
2) Remote support for disabled persons with guidance
3) Mobility support
4) Road security
5) Mobile office

They will address the design of new systems and terminals even beyond 2050.

13.4.1 Remote Support for Elder and Diseased Persons

One of the most important foreseen needs in the next decades is related to health. This is due to a longer life expectation and the need of persons to monitor their health status related to their age. Many different health sensors will appear to widen the illnesses control as what happens today for diabetes. A sensor measures blood's glucose concentration to activate a small pump to inject continuously the right quantity of insulin. Moreover, acquired data are processed by health experts, and appropriate actions can be taken in order to keep the person in his/her best health state. Other APPs could suggest people to perform physical activity following a prescribed schedule in order to maintain an appropriated movement fit and motor ability. Physical exercises are tailored to the person and help them to be active. They could consider the natural physical activity done during the day, for example, walking at home or outside, muscle strengthening doing domestic job, or gardening, and propose to complete the daily training by using the virtual reality (VR) system. VR system is actually based on either a TV set or a large monitor where to display games to involve elderly into activities. With the increase of link capacity, we could transfer game from remote directly to the display, which could be a standard TV set, the smartphone display, or a screen embedded in eyeglasses as suggested before. Then, during the next decade we will see a completely different setup for VR system, and games could be played even in outer space as public garden.

For different and complementary activities, this new application could allow elderly to maintain their independence in a good health state during their life. This would reduce costs for medicines and medical checkup. Moreover, the system could easily embed videoconference, which would allow elderly to keep in contact with friends even in the case of low mobility. This aspect surely would improve the perceived quality of life.

Other applications will be developed in the future, and their success depends on their easy-to-use mode and on the special needs they are considering. They could be easily included in the home system composed of a PC, a TV set, some local wireless links, and a broadband access to Internet and extended in the near future to smartphones and augmented vision systems by using active eyeglasses.

13.4.2 Remote Support for Disabled Persons with Guidance

The number of disabled people is increasing with the average age of the world population. ICT is a powerful means to help these people to be autonomous and to live an independent life. Solutions exist even today, but they are very expensive, and only a small number of persons who need them are able to get. Again, mass production of these objects will reduce their costs and their availability to almost all who need them. Disabilities are of several different natures but they can be separated into few classes, which ICT can easily match to: (i) mobility limitation, (ii) blindness, and (iii) cognitive disability.

13.4.2.1 Mobility Limitation

People with limited mobility ask for three different supports: (i) physiotherapy exercises to regain proper mobility of legs and arms, (ii) exoskeletons to support person with a proper mobility, and (iii) electric carts to move around to overcome paraplegia. ICT could bring advantages in all these cases. Several studies are ongoing today and in one decade they will provide products available for large-scale use. Next, evaluation of their efficacy will be needed, and improvement of their performance in several different clinical cases is expected.

Physiotherapy Exercises to Regain Proper Mobility of Legs and Arms VR systems can be used to design games that gradually bring patients to perform customized exercises guided by expert physiotherapists. These exercises could be done at home, and the hardware needed is limited to a PC and a digital TV set. The design of games could be done even by small enterprises, which are connected to physiotherapists for understanding how to organize the movements embedded into the games. It has to be remarked that, in order for these games to be used everywhere, a well-defined standard to operate the game based on the human joint model [22, 23] is needed. A known standard composed of 48 joints already exists: however, this model is not sufficient to cover all the rehabilitation procedures, which require fine movements of all parts of human body. Therefore, the idea is to start from the available model leaving it open to further definition of new nodes to be used when needed.

Nevertheless, the existing model has to be refined as well by introducing limitations to the mobility range and the force applied correlated with each specific joint.

Exoskeleton to Support Person with a Proper Mobility Exoskeleton is used today to help people to move alone in some cases. In such a way, patients are able to perform movement with many sensors applied to their body recording and controlling all activities. Recent researches look into using exoskeletons or robotic devices, which are controlled by EEG/EMG signals to drive limb rehabilitation protocols. Performance of these devices is encouraging, and new devices are on the way to be designed to further improve to even more successful results. In this way the machines, which need several control signals, would be based on multidisciplinary research results related to neurological studies, physiotherapy knowledge, new exercises, and ICT system design in order to perform all needed signal processing. Wireless communication in this field helps a lot because it would enable to drop out wires, which typically connect EEG caps and all the sensors applied on the person's body.

Electric Cart to Move Around   In case of severe disability, people are not able to move alone even with the support of exoskeleton. In this case mobility could be guaranteed by an electric cart. An autonomous system to drive safely the cart along the way is still lacking. This aspect requires many ICT techniques, some of them already available, some others still under investigation. Most of the techniques require wireless communication to allow for safe transportation. Here, some of the technology critical elements are listed: a positioning system to know where the cart is, a short-range radar system to detect obstacles around the cart and avoid them, an intelligent system to interpret the command from the person driving the cart, and so on. Some of these devices are on the way to be tested in a road environment https://www.google.com/selfdrivingcar.

In conclusion, today's research in the laboratory is producing several different ICT support technologies for the personal mobility. By 2050 these devices are expected to be further improved in order to be regularly used for maintaining elder people either in good physical health or to help them to maintain their mobility independence.

13.4.2.2 Blindness

Blind people need ICT help to move safely along the street. Today there are few technologies helping them to move: indeed most of the people are still helped by Seeing Eye dogs to walk. In few years ICT solutions for self-driving cars (see Section 13.4.4) could be modified in order to be used to guide safely the walk of blind people. Applied research in order to generate clear commands and to design a lightweight, low consumption, easy-to-wear system is mandatory. Therefore, by 2050 we expect that ICT will enable blind individuals to move safely alone everywhere.

13.4.2.3 Cognitive Disability

Cognition refers to the processes by which the sensory input are transformed and used. Cognitive disability is then related to the impairment of performing some mental processes like understanding speech or producing correct words; paying attention, that is, to focus on a subset of received perceptual information; performing logical and/or mathematical operations; and so on. Cognitive rehabilitation is a long process in which the patient is trained to improve his/her mental processes. This is typically done by administering attention exercises or by employing speech therapy. Both therapies could be performed by using a PC and a touch screen. Exercises to improve attention are based on the submission of sheets where a symbol has to be searched or mathematical logical operations with multiple solution choices have to be solved. These exercises could be easily implemented by a PC with a touch screen monitor where the patient can touch the guessed solution.

For speech rehabilitation words/sentences are pronounced and the patient has to repeat them. The system provides a feedback concerning the correctness of the pronunciation of the word/sentence ask to the patient.

This system is today available but could be improved a lot in order to be more effective. Joint team of researchers composed of engineers, psychologists, and neurologists could do the job, and we expect to have better performance by 2050.

13.4.3 Mobility Support

Navigation systems identify the position where we are in a precise way and give us indication for the route to follow in order to reach a known destination. A data base could integrate this basic information with others tailored to the person needs. A tourist represents an interesting example: he/she is moving around and asks for getting information about history of the site, museums, and famous masterpieces. Data bases contain this information and deliver it to the client in an appropriated format. The wireless part is limited to the connection and to display images and maps on the mobile screen.

In order to satisfy client needs, data bases have to completely cover different areas, for example, museums, restaurants, hotels, scenic areas, and playing areas. The way in which information is given to the clients could depend on their terminal. Moreover, the information could be either a short summary or a detailed description. The organization of the data bases and how to display the information will continuously evolve following the client requirements.

13.4.4 Road Security

The most intriguing project on road security is the *Google self-driving car project* (https://www.waymo.com), whose purpose is to design a car that autonomously bring us to the final destination moving within standard road with standard traffic. The project began in 2004 when DARPA organized the first "grand challenge," a fully autonomous vehicle competition run. No vehicle concluded the first competition run, but the year after five vehicles completed the 240 km run. The first four teams ended the run with a time varying from about 7 h to 7 h 30 min, while the fifth arrived after about 13 h. The next grand challenge was organized in 2007 and was devoted to an urban run: 96 km to be completed within 6 h. Seven teams completed the run; the best spanned 4 h and 10 min to arrive.

Next, Google is pushing forward its self-driving car ambitious project. Its next purpose is to reduce the dead counts due to road accidents, which is extremely high around the world. The technology to support the self-driving vehicles requires a lot of ICT that includes, among others, a vision system to detect the road lane, other vehicles and moving people, street signals, and a modified radar device to detect nearby objects. A short-range communication system is used to communicate with the neighbor vehicles in order to inform them about the direction of motion and gather similar information from them. In this way it would be easier to decide what to do. Street signals could be detected either from the signal deployed along the streets or by signals broadcasted from the road infrastructure.

From this overview within the next 20 years, 20% fully autonomous new cars could be expected. In the most advanced countries, roads could have signals, which will be broadcasted, and, along the road, stores will advertise themselves on a dedicated channel. Then, the car passengers could organize a detour to visit stores or interesting buildings or places.

In this new scenario the transmission of information to the final user will likely require new services.

In this way the road will be much more secure than today, and the passengers could enjoy knowing better the places they are crossing by, thanks to the availability of information (historical, monumental, business, etc.) of the site they are moving around.

13.4.5 Mobile Office

Nowadays, the opportunity to be connected with the office networks everywhere allows workers to work at home without needing to be physically at the working place. Moreover, automated factories could be managed from remote. As a consequence, workers could perform their job everywhere in time and space. Globalization of the job market has started to take into account different time zones of the world: thus, the standard job time 8 am–noon plus 1–5 pm sounds old. These two aspects are bringing workers to organize their job time taking care of their personal as well as work-related needs. In this way the possibility to perform the job everywhere and every time is something new and will be considered in the future.

The evolution of this aspect of the job is still beyond a reliable perspective because it is mainly related to many different legal and social aspects still undebated.

13.5 Conclusions

As we discussed earlier, ICT is close to a new revolution that is expected to peak by 2050. While in the past century the technology was the driving force for the deploying of new systems, nowadays needs of clients trigger the technological progress. This fact will change completely the development of new services, and, in some sense, it appears as a democratic way to access technology. This new paradigm will change several aspects of the daily life even for engineers. Till today engineers were oriented to design new machines based on their expectations about the future. However, now on, clients will ask for needs and engineers will have to find the way to design services that satisfy them. This fact will create new interactions among people with different backgrounds as well as languages: thus, a basic culture and a common language will be needed.

Another important aspect is how social relationships will be shaped by wireless technology: young people easily meet by using chat line, social media, and so on and are reducing the time they spent in a *vis-a-vis* meeting. This is creating a society where people virtually exchange experiences, information, and even emotions: a new revolution. Furthermore, augmented reality and VR will become part of our future lives and will help us by limiting at the same time the opportunity of interaction with other people. These new kind of relationships will probably largely influence our living and have already become a hot topic of psychology, sociology, and philosophy.

References

1 W. Shockley, 'The theory of p-n junctions in semiconductors and p-n junction transistors', *Bell System Technical Journal*, vol. **28**; pp. 435–489 (1949).

2 J.S. Bardeen, and W.H. Brattain, Three-electrode circuit element utilizing semiconductive materials, US Patent 2,524,035, Bell Telephone Labs, Oct. 3, 1950.

3 W. Shockley, Circuits element utilizing semiconductive material, US Patent 2,569,347, Bell Telephone Labs, Sept. 25, 1951.

4 J.S. Kilby, Miniature semiconductor integrated circuit, US Patent 3,115,581, Texas Instruments, Dec. 24, 1963.

5 R.S. Noyce, Semiconductor device and lead structure, US Patent 2,981,877, Fairchild Semiconductor, Jul. 30, 1959.

6 B.M. Oliver, J.R. Pierce, and C.E. Shannon, 'The philosophy of PCM', *Proceedings of the IEEE*, vol. **36**; pp. 1324–1331 (1948).

7 K. Odaka, Y. Sako, I. Iwamoto, T. Doi (all from Sony), and L. Vries (Philips), Error correctable data transmission method, US Patent 4,413,340, 1983.

8 K. Immink, J. Nijboer (both from Philips), H. Ogawa, and K. Odaka (both from Sony), Method of coding binary data, US Patent 4,501,000, 1985.

9 W. Sweet, 'Chiariglione and the birth of MPEG', *IEEE Spectrum*, vol. **34**; pp. 70–77 (1997).

10 G.H. Boone, Computing system CPU, US Patent 3,757,306, Sept. 4, 1973.

11 F. Faggin, M.E. Hoff Jr., S. Mazor, and M. Shima, 'The history of the 4004', *IEEE Micro*, vol. **16**; pp. 10–20 (1996).

12 G.O. Squier, 'Multiplex telephony and telegraphy by means of electric waves guided by wires', *AIEE Transactions*, vol. **30**, pt. II; pp. 1617–1665; discussion, pp. 1666–1680 (1911).

13 E.H. Colpitts, and O.B. Blackwell, 'Carrier current telephony and telegraphy', *Journal of the AIEE*, vol. **40**; pp. 301–315; pp. 410–421; pp. 519–526 (1921).

14 H. Nyquist, 'Certain factors affecting telegraph speed', *Bell System Technical Journal*, vol. **3**; pp. 324–346 (1924).

15 H. Nyquist, 'Regeneration theory', *Bell System Technical Journal*, vol. **11**; pp. 126–147 (1932).

16 H. Reeves, Electric signaling, French Patent 852,185, Oct. 3, 1938; US Patent 2,272,070, Feb. 3, 1942.

17 C.E. Shannon, 'A mathematical theory of communication', *Bell System Technical Journal*, vol. **27**; pp. 379–423; pp. 623–656 (1948).

18 H. Cravis, and T.V. Crater, 'Engineering of T1 carrier system repeatered lines', *Bell System Technical Journal*, vol. **42**; pp. 431–486 (1963).

19 W. Rae Young, 'AMPS: introduction, background, and objectives', *Bell System Technical Journal*, vol. **58**; pp. 1–14 (1979).

20 V.H. Mac Donald, 'Advance mobile phone service: the cellular concept', *Bell System Technical Journal*, vol. **58**; pp. 15–41 (1979).

21 Z.C. Fluhr, and P.T. Porter, 'AMPS: control architecture', *Bell System Technical Journal*, vol. **58**; pp. 43–69 (1979).

22 S. Chattopadhyay, S.M. Bhandarkar, and K. Li, 'Human motion capture data compression by model-based indexing: a power aware approach', *IEEE Transactions on Visualization and Computer Graphics*, vol. **13**; pp. 5–14 (2007).

23 J.M. Baydal-Bertomeu, J.V. Durà-Gil, A. Piérola-Orcero, E. Parrilla Bernabé, A. Ballester, and S. Alemany-Munt, 'A PCA-based bio-motion generator to synthesize new pattern of human running', *PeerJ Computer Science*, vol. **2:e102**; https://doi.org/10.7717/peerj-cs.102 (2016).

14

Human Bond Communication Beyond 2050

Flemming Hynkemejer[1] and Sudhir Dixit[2,3]

[1] RTX A/S, Wireless Wisdom, Norresundby, Denmark
[2] CTIF Global Capsule (CGC), Rome, Italy
[3] Basic Internet Foundation, Oslo, Norway

14.1 Introduction

This chapter discusses the paradox of people on one hand being more and more into communication platforms and applications, which should provide all the opportunities in the world to exchange information and data in increasingly efficient ways, and on the other hand the choices made by us as telecom customers for the platforms that only use the least significant parts of the messages (i.e., text). Studies show that only about 7% of the total information is included in the text, whereas sound (intonation, volume, speed tonality, etc.) and video (body language, eyes, etc.) convey 38 and 55%, respectively. The other three human senses (touch, smell, and taste) are completely excluded.

Why is it that especially the younger generations communicate increasingly more through a "low quality" media than face to face, and *Short Message Services* (SMS) prevail over face time or even voice, when we know that so much information is lost in translation (or limited channels)? How will this influence our societies in the long term?

The questions we need to answer are as follows:

1) Why do we tend to choose poor(er) quality platforms for communication? (Is it price, convenience, or human temperament?)
2) Will people avoid communicating face to face if given the choice? And if so why?
3) Would people eventually lose the ability to read body language or is it part of the DNA? If it can be lost, how many generations would it take?
4) What will it mean to society/individuals if we are not communicating face to face?

Human Bond Communication: The Holy Grail of Holistic Communication and Immersive Experience, First Edition. Edited by Sudhir Dixit and Ramjee Prasad.
© 2017 John Wiley & Sons, Inc. Published 2017 by John Wiley & Sons, Inc.

The chapter will not provide answers but merely address the issues and hopefully motivate the research community and industry to seek answers to these questions.

14.2 Origin of Communication

One cannot doubt that language owes its origin to the imitation and modification, aided by signs and gestures, of various natural sounds, the voices of other animals, and man's own instinctive cries [1].

Since the dawn of days, humans have communicated by using gestures formed by the body, that is, facial muscles, body language, hands, movements, and sound for information transmission and senses, for example, eyes, sense of touch, nose, and ears for reception.

It is not clear when man developed language but it seems logical that language was an add-on to gestures and movements. Some scientists suggest that the need for man to evolve a language came from the need of using hands elsewhere, for example, holding tools and so on, or from not being in visual contact. Both hypotheses are based on a desire for increased flexibility.

The human culture and evolution is based on passing stories, experience, and skills to the next generation. To create a situation, an atmosphere, and channel for passing messages, humans needed two elements—fire and language. In the olden times telling stories around the fire was the basic entertainment available and humans needed entertainment to assess their feelings; feelings were fuel for breeding and breeding was fuel for evolution. The stories told were long lived and thus essential for human existence.

Even though the vast majority of communication today is language either in the form of written letters or spoken messages, the value of processing the

Figure 14.1 Origins of communication among humans.

information contained in the unspoken part of the communication must not be underestimated. Evolution has given us language but brought fragments of the unspoken communication forward as significant parts of the total data information.

Depending on the individual, visual, auditory, or kinesthetic stimuli will prevail as the preferred channel for receiving and processing information. Albert Mehrabian's studies in interpersonal communication show that only about 7% of the total information is included in the words, whereas audio (intonation, volume, speed, tonality, etc.) and visual (body language, eyes, etc.) stimuli convey 93% of the total information; see Figure 14.2.

This study has been misinterpreted many times, as it suggests that one only needs to see and hear the communication to understand more than 90% of the total message (even in a foreign language that is not understood).

However, the context of the study was meant to decide whether a person had preferences (likes) for this or that:

$$\text{Total liking} = 7\%\,\text{verbal liking} + 38\%\,\text{vocal liking} + 55\%\,\text{facial liking}$$

Thus the study merely reflects the communication channel, and the communicator is talking about his/her feelings or attitudes, or else this equation is not applicable.

Nevertheless, it is true that there is important information hidden in the nonverbal part of face-to-face communication, and sometimes we are better and faster at reading nonverbal signs than the actual spoken words.

In modern life, we have supplemented face-to-face communication with a set of alternatives such as voice-to-ear communication, old (long) handwritten letters, short messages (e.g., SMS), and video/photo sharing like Instagram, Flickr, and so on.

Other studies, like Mehrabian's 1971 study, map the telephone channel as the face-to-face channel and show, not surprisingly, that words become

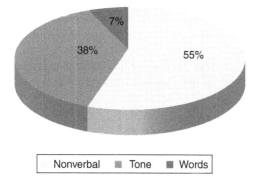

Figure 14.2 Mehrabian's 1971 study on face-to-face communication.

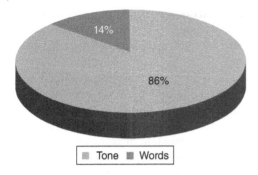

Figure 14.3 Model for telephone communication.

somewhat more important, but still a lot of information is conveyed in the tonality; see Figure 14.3.

Again this study should be seen in context, but the key question is whether the availability of alternative communication platforms has improved human skills in communication and interpersonal relations. One would argue that these studies have ignored (not necessarily by design) the value of smell, touch, and taste for complete human experience. This aspect probably did not occur to them, but it is presumably going to change with the advances in technology and the need to innovate in an increasingly commercialized and competitive world.

14.3 Technology as Enabler for Communication Improvement

Telecommunications engineers for centuries have strived to improve communication between people, by bringing products and applications to the market that allowed them to communicate in many ways. Samuel Morse's invention of the telegraph in the 1830s allowed the average man to send a message across the world despite a high price.[1] The costs of early telegrams were only justified by the value of instant communication.

The main challenges for the telegraph were (i) cost and (ii) a limited message capacity, so when Graham Bell invented the telephone in 1880, customers were

---

1 In 1860, for example, a 10-word telegram sent from New York to New Orleans cost $2.70 (about $65 in 2012 currency), while a 10-word transatlantic message to England cost $100 (about $2600). These prices declined in time, but telegrams largely remained a tool for the corporate, the rich, and for emergencies. Despite the high cost, some 212 million telegrams were sent to the United States in 1929, the peak year for such activity. Source: http://www.voanews.com/content/india-marks-end-of-era-with-last-telegram/1701749.html.

offered to communicate in real time over long distances in their voice and tonality, and thus a great leap happened in worldwide communication. But at the same time, the telephone users or telegraphers were needed to be present at the same physical location as the telephone/telegraph apparatus, and the recipient or operator also needed to be at the exact location and time, where the sender would place his or her call. In addition, messages or calls were managed by a local operator. The advantages of having a personal operator were many, for example, he or she could take a message or give notice if the recipient was busy on the line and even interrupt a call if necessary. Later, the switchboards were automated to save cost, and many years and billions of dollars were subsequently spent to replicate this personal operator function, for example, switching calls, voice mail box, notification if the recipient is occupied or busy, and even personalized ringtones.

Since the 1980s advances in technologies have fueled mobile communication platforms and profoundly changed our way of communicating, we are now able to communicate personally to anyone from anywhere at anytime in multimedia—text, audio, and video.

We have gone from fixed line short-letter messages with educated recipients' presence (Morse code operated telegraph) to SMS[2] to image sharing to a place where more people tend to document their complete life in the cloud. Similarly, engineers orchestrated a path over voice (POTS) to mobile voice and video conference platforms. And the reality today is that people now communicate on various platforms at various times and may even change platforms several times during a single conversation.

In the 1920s radio broadcast became popular, and for many years these broadcasters were strictly regulated and controlled by governments and authorities. In the 1950s broadcasters included video to the palette and suddenly we had television. The setup was given and the regulator determined what content could be transmitted and at what time.

Until the mid-1990s things were fairly stable despite having more TV channels and a more liberal approach, but as more efficient transmission technologies became available, the business models for content started to change. In fact, in the beginning, content was basically made available for free, and the only really costly content was big sports events as the Olympic Games, World Cup in soccer, and other sports. Soon thereafter, content became king with price attached to it. After the Internet revolution, everyone was on the web, and for a decade we have been able to create and share our own individual media/content to the whole world through Facebook, Instagram, Flickr, and so on.

---

2 The market adoption of SMS was relatively slow, probably because telecom operators did not believe in the business and thus did not rush to set up billing systems for the service for a long time. The first SMS was sent in 1985 and just in the late 1990s, the service was broadly adopted.

We are now in an era where consumer is king, a time of anti-Copernicanism[3] where for one tiny bit, man is (again) the center of the universe.

Especially on Facebook and blogs, where users have a well-defined profile, the ability to recommend, promote themselves, and sell products have made it easier to target specific individual customers with specific and relevant marketing material. The full scale of this has probably not developed yet, and the interim steps include bloggers and trendsetters using themselves as promoters for various products and services.

Trying to understand the basic differences of these communication platforms and why customers seem not to choose the channel that offers optimum communication in an objective sense, let's discuss the current trends:

- From POTS to *mobile*. During this transition, technology offers users *flexibility in place* while still replicating the established voice-to-voice experience.
- From telegraph to *SMS* (SMS, MMS, Twitter, etc.). Understanding that the last telegraph operation was shut down in 2013 (in India) may link the two platforms closer together after all. *Cost of service* is definitely one of the success factors, and SMS has most likely succeeded in attracting many users because the technology is cheap. The concept of *short convenient messages or images that support or prolong a longer dialogue* seems to be used more than the channel itself could attract.
- From radio broadcast to be your own media producer. Content anarchy is creating *personal images or content broadcasted from the creator* directly to everyone on the network, and not necessarily to a specific recipient. YouTube, Snapchat, and Flickr are platforms where people can communicate in a one-to-many dialogue using photos, videos, and text. Depending on which channel, *some of these services are time restricted* so that you are only able to access content for a limited amount of time after which content is destroyed. This enforces users to be online and regularly check for updates. It also allows adding content to the dialogue that is not necessarily intended to be recaptured at a later point in time.
- From mass marketing to needle pin marketing via Facebook and Instagram. The *high-end* social media with most subscribers[4] *allow people to share their life stories* across the globe. Being able to share our private episodes with only the closed network, we are now able to *push* our stories across a broad and

---

3 Anti-Copernicanism—Martin Luther (1483–1546) is alleged to have criticized Nicolaus Copernicus (1473–1543) for disturbing science with his heliocentric system, thereby the name anti-Copernicanism.
4 Monthly active users (MAUs) were 1.44 billion as of March 31, 2015. An increase of 13%, year over year. Source: Facebook quarterly report March 2015.

loosely defined network. Those with most followers[5] and *most interesting profiles are able to market products and ideas* to their followers.

Now we have established a superficial understanding of how technology has influenced communication:

1) Offer access everywhere (impulse, accessibility, and convenience oriented)
2) Given an option to communicate to the end point over concise (uninterruptable and efficient) channel
3) Attractive cost of service (affordability)
4) Provide the ability to speak without being forced to reply immediately (emotion)
5) Ability to share information that is meant for a context and not intended to be cited back (impulse and emotion)
6) Personalize content that is not the same as relevant content but enables the individual to present a *retouched* image and to receive *angled* information addressed to the person (filter)
7) Preprocess (and post-edit) content where the recipient allows friends and network to filter and process marketing material for products and services (filter)

14.4 Building a Basket of Communication Platforms

An opportunity given is often an opportunity taken. The more platforms for communication available also mean that individuals are inclined to communicate at anytime to anyone, anywhere. They can also decide to whom they want to listen to, how to filter information, and when to listen.

How can we get a better comprehension of how a user decides which communication channel is optimum for getting their message across?

To bring our understanding a step deeper, let us consider a dialogue between two parties A and B. A is communicating a message to B, and B replies back. Let us investigate how this is experienced on various channels, and we may understand why different platforms are used.

The parameters could be numerous, but let's just consider speed of communication, distance, physical constraints, message reception, and options for replication (See the table in the next page).

It is easily spotted that the more advanced the platforms become, the more fulfillment of ambitions is experienced (more green). It is also observed that

---

5 On September 14, 2011, Facebook added the ability for users to provide a "Subscribe" button on their page, which allows other users to subscribe to their public postings without needing to add them as a friend. See: http://mashable.com/2011/09/15/facebook-subscribe-users/#YyIG6SAaxOkh.

Different communication channels and technical characteristics.

Platform	Latency from A sends message to B receives	Distance/ reachability	A and B physical constraints time and place	B direct influence on A message	B understanding (theoretical channel bandwidth) (%)	B replication
Face to face	Real time	Within a few meters	Locked in time and place	High—interruption and nonverbal	100	Immediate 100% channel
Letter	Days	Basically anywhere fixed	No constraints[a]	None	>7	Delayed with days
Telegraph	Hours	Post office or delivered	No constraints	None	7	Delayed with hours
Fixed voice to fixed voice	Real time	Basically anywhere fixed	Locked in time and place	Medium— interruption	45	Immediate but limited channel 45%
Mobile voice to mobile voice	Real time	Basically anywhere	Locked in time	Medium— interruption	45	Immediate but limited channel 45%
SMS/MMS/ Instagram/e-mail	Seconds (almost real time)	Basically anywhere	No constraints	None	7–100[b]	Seconds but limited channel 7–100%
TV broadcast	Real time	Basically anywhere fixed	Somewhat locked in time and place	None	100	None—one way communication
YouTube	No certainty for delivery	Basically anywhere	No constraints	None	100	None—one way communication
Mass marketing	No certainty for delivery	Basically anywhere	No constraints	None	100	None—one way communication
Facebook	Real time, if online	Basically anywhere	No constraints	None	100	Easy, but not necessarily intended

Light and medium grey indicates more fulfillment of experience, dark grey indicates least fulfillment, and white indicates moderate fulfillment.

[a] No constraints except that you need a post box/office for sending and an address (mailbox) to receive a letter.

[b] Including attached files, photos, and so on.

the early electronic communication platforms were doomed as new platforms improved initial latency time, reachability, channel, and/or removed physical constraints, that is, there was no going back once some obstacles were removed. But still, some of the new platforms are seen as add-ons and not substitutes to older ones, for example, we still use mobile while we could solely rely on SMS and MMS. It should be noted that the substitution effect is difficult to measure and no adequate studies have been done so far. Later we will see how much consumption there is on various platforms and where it appears that newer platforms are more complementary compared to substitution.

An interesting aspect of the table is that the lower end with both newer social media and time-consuming media as TV (~3h consumption per day per inhabitant in Denmark in 2015) tends not to target dialogue but to focus more on monologue.

On the other hand, the *bandwidth*-limited channels offer another important aspect of communication—the psychological aspects of a conversation—the ability to speak in due time or communicate things not to be said face to face or the ability to subsequently regret parts of a conversation.

14.5 Psychological Influence on Communication

Throughout history many strong relations have started from letters and written dialogue and not so much from face-to-face encounters. This supports the argument that it is not always the broadest and most advanced channel for communication that ends up delivering the best message exchange platform.

Having the time to think about a dialogue, formulate precise responses, and be exact and concise have often added value, quality, and even efficiency to a conversation. On the other hand, misunderstandings could take weeks to solve. The point is that time (i.e., speed of conversation), not being a constraint, may compensate for limited visual and aural bandwidth. Having the opportunity to reread the conversation over and over again can also strengthen relationships and conversation quality, compared to high pace and confronting face-to-face encounters.

In recent years there has been some public discussion of what to say and what *not* to say, for example, on the SMS medium. Is it okay to fire an employee via SMS? Is it okay to report in as sick on SMS? Is it okay to drop a sweetheart on SMS? These are interesting scenarios because these non-face-to-face platforms allow people to deliver unpleasant messages while hiding, via short text messages without being questioned for their motives and not allowing any discussion and context into the dialogue.

Another aspect is the ability to post-edit a conversation. The old-school letter was per definition sent as it was, so was the telegram, SMS, and e-mail.

However, when it comes to social media, the editing limits become more blurred and allow the editor to post-edit or even remove information entirely subsequently. This allows a sender to change, improve, or discard information shared earlier in a conversation.

In Table 14.1 a comparison between various platforms is made with respect to investigating psychological proportions of the channel, that is, speed of dialogue. Is the recipient able to access the message over again? Will the sender be able to reframe the message after initial transmission, and lastly will the two parties be able to communicate without confronting the other party directly?

Some people are introvert as others are extrovert, and in many years it has been *comme il faut* to train introverts to become more extroverts. The availability of social media has made it possible to address networks with only a virtual presence, thereby allowing people to hide behind the screen and only interact through a keyboard and/or camera. This may bring introverts a new opportunity to bring themselves onto the scene despite being timid. On the other hand, those who are most popular in the real world (typically extroverts) will tend to take a stronger position in the social media, thereby creating an even bigger polarization between the introverts and extroverts.

This may set humankind backward as we now can end up in a situation where society becomes generally more introvert as the bright light will only

Table 14.1 Psychological aspects of communication.

Platform	Time constraints for response	Ability to reread message	Post-editing ability	Ability to communicate non-confronting
Face to face	High pace	None	None	No
Letter	Low pace	Yes	None	Yes
Telegraph	Low pace	Yes	None	Yes
Fixed voice to fixed voice	High pace	None	None	No
Mobile voice to mobile voice	High pace	None	None	No
SMS/MMS/ Instagram/e-mail	Medium/ low pace	Mostly yes	None	SMS/MMS/e-mail no Instagram yes
TV broadcast		Some	Some	
YouTube		Yes	Yes	Yes
Mass marketing		Yes	Some	Yes
Facebook	Medium/ low pace	Yes	Yes	Yes

shine on the few media darlings compared with the gray mass of introverts, but now not being forced to train their underperforming extrovert capabilities. A counterargument could be that instead, social media allows introvert people to communicate and thereby overcome their introvert profile. The reality will probably prove both arguments valid. Conclusion is that different platforms adapt to different psychological preferences, and thus the choice of channel becomes even more complex. The choice of platform may thus depend on many variables, for example, sender's psychological profile, the message, and the expected response.

Let us now look into the actual consumption of the various choices just discussed.

14.6 Platform Consumption

This section discusses that given the choice, people seem to create their own basket of communication services, but there are some general trends:

1) Old-school platforms, such as POTS, mobile, and SMS, seem to be long lived and rather stable in consumption.
2) New social media platforms tend to focus more on monologue than dialogue.
3) New social media is not about communicating real time but about being online 24/7.
4) There may be a demand for the ability to reread information.
5) There may be a trend for low pace communication platforms.
6) Social media allows post-edition.

The next step is to investigate the actual consumption to understand whether the options given are options taken. Table 14.2 shows consumption on Danish domestic communication platforms based on statistical figures from the first half of 2014.

Whereas the consumption of speech (minutes per inhabitant) and SMS have been fairly stable over time and speech consumption seems to show small signs of decrease, it is interesting to see that despite stable voice consumption, the new communication platforms have added on top of the old platforms.

TV watching, which is one of the big time consumers, has been weakly increasing over the past 10 years even though the annual statistics are interpreted by journalists claiming that now babies are watching less TV and the old are watching more, and next year vice versa.

As seen in Table 14.3, a substantial part of Danish inhabitants is now active on the social media. Even though some are more users by name than actual use (monthly vs. daily users), the trend is clear—users tend to supply their communication basket with social media.

Table 14.2 Domestic telco infrastructure and high-level communication patterns (Danmarks Statistik Telerapport 1H 2014).

Denmark: 1H 2014		Development year on year
Inhabitants	5.64 M	Weakly increase
Mobile speech subscriptions	7016 M	Stable
Domestic traffic minutes (mobile and fixed)	6142 B	Weakly decrease
Traffic per inhabitant	6 min/day	
SMS	4421 B	Stable
SMS per inhabitant per day	3–4 messages/day	
TV consumption	173 min/day[a]	Stable
YouTube consumption	6 min/day[b]	NA

[a] 2014 Yearly average. Source: Danmarks Statistik.
[b] January–September 2014 data. Source: DR Medieforskning.

Table 14.3 Take-up rates of social media.

Social media penetration (2H 2014 or see footnotes)	Denmark (penetration relative to population) (%)	Worldwide adoption rate (million users)
Facebook (daily users)	~70	1440
YouTube (monthly users)	~40	1000[a]
LinkedIn (monthly users)	~22	347[b]
Snapchat (monthly users)	~22	200[c]
Google (monthly users)	~21	540[d]
Instagram	~17	300
Twitter (monthly users)	~12	316

[a] Source: 2013 figure from YouTube.
[b] Source: 2015 figure from LinkedIn.
[c] Source: September 2015 figure from theverge.com.
[d] Source: USA Today.

The question is now whether TV + social media is a supplement or a complement to face-to-face meetings. Many studies show that people are feeling increasingly lonely. A study from spring 2015 shows that in Denmark 210,000 persons aged 16+ years feel alone often or all the time. This corresponds to approximately 5% of the population. This may support the argument that technology no longer bridges the gap between people but instead increases distance.

If we assume that users tend to spend time on the top three penetrating social media (Facebook, YouTube, and LinkedIn) either to edit their own story/

image or to look at others' images/stories, then this further supports the idea that communication has not increased in the form of dialogue, merely the opposite. However, the social media has opened new ways of sharing stories.

The list of conclusions is long:

1) People are still living in the Stone Age and like to have a good story told around the fire. Now the TV acts as the bonfire and is used as the most important storyteller. TV is still the main time-consuming media with approximately 3 h/day/capita (4+ years).
2) New services with success tend to be mobile on-demand services, allowing people to access them freely in time and space.
3) On average Danish people spend as much time on YouTube as they do on the voice-to-voice services.
4) Social media tends to add on top of traditional media, thus taking time from other things.
5) The most penetrated social media tend to create distance compared to balanced dialogue.

This implies that humans when given the choice are likely directed toward solitude media consumption instead of face-to-face dialogue, or it may indicate that this media consumption lies in the edge of people' daily programs, and therefore need to be on-demand and easy to access/digest.

14.7 What Is Next?

Content will continue to be the king, and users would play an increasing role in creating, communicating, and consuming over the democratized Internet. This content would be richer with audio, video, text, and augmented reality, all combined in a digital world, customized to user's individual profile and the context. The characteristics of the subject, whatever it may be (person, thing, nature, etc.), will possibly not only be characterized and rendered through audial and visual means but also be augmented through all three (or a subset of) remaining senses—smell, touch, and taste—depending on the use case and the transmitter and receiver setups. This will, eventually, make the world of communication go full circle to historic old times of face-to-face communication and storytelling, albeit digitally over long distances and shrinking the world across societies and cultures.

14.8 Conclusions

This chapter has tried to understand some implications that technology has had on communication and how humans have adapted to it over time.

We presented some ideas illustrating (through storytelling) how communication was essential for developing the human culture and the civilization. In the (good) old days, stories tended to live long and possessed certain qualities for securing evolution and pace of society.

This conclusion section discusses a link between the past, the present, and the future by looking into some of the mega trends that we have witnessed in the recent years. We have found that people will continue to satisfy their demand in a context evolving from quantity to quality and from complexity to simplicity, mainly because of the dwindling resources on our planet Earth and as a general response to ever increasing complexity.

With respect to communication the development has been in the opposite direction, as a result of the exploding content streaming over the Internet and the popularity of social media.

What will be the nature of our society when stories are completely individual, when there are billions of small bonfires where stories are told maybe, even more than one per inhabitant of the globe, in an environment where you can travel from one fire to the next in a split of a second?

How can we tap this almost inevitably large source of intelligence and experience? This still remains a challenge, but lots of new opportunities have sprung up through research and development in big data and analytics. The technological (and/or more importantly anthropological) challenge is to understand and develop tools that allow us to derive and sort out valuable information from this vast amount of data and to extract essence and meaning to accomplish further development and evolution of the humanity. If this becomes possible, our history will be judged in the way we process data and not the data itself. In ancient times, data and stories were used for documentation, tradition, and basis for evolution; now we tend to drown ourselves in data, and thus we have to use data for monitoring and controlling *on the fly*. What implications will this have on our ability to understand and create context is an open question?

There is probably going to be some polarization among those who can maneuver (with fast response time and high data processing capacity), giving them advantage, and those who just submerge in data.

We discussed how nonverbal communication conveys important parts of a message, if the message is interpersonal and relates to feelings and preferences. The augmentation of aural and visual communication with the three remaining senses (tactile, olfactory, and gustatory) will further enrich the experience. This, of course, comes at a price of sacrificing freedom in space and wireless accessibility. Some attempts to overcome this have been made through the usage of emoticons, for example, smileys, but a lot of work still remains to be done.

To be able to respond to the increasing amount of data, we will need more efficient ways to communicate on the same level of convenience as available today. Using a basket of communication platforms makes it possible to match the message to the person(s) intended with the optimum communication channel. We hypothesize that the recipient of the message should decide on the quality of the communication. We have established an understanding supporting that the channel is also part of the total communication, and thus future communication platforms may allow automatic channel shifts, depending on the content, for example, translating spoken words into written text and simulcast text and voice.

With an increasing population that at the same time feels increasingly more alone than ever before, leaders of societies will need to take action. The desire to have stories told remains and will probably remain for the time to come. The content of the stories seems to be the issue. TV will only act as the ancient bonfire if the content is relevant and brings people together (virtually speaking). Without content (stories) that are able to bring people together, the basic foundation of societies will eventually disappear. This may open for new societies based on virtual reality societies where physical and geographical boarders are abandoned and people live more in smaller communities or tribes as in the historical past. The communication platform will then have to form and provide the basic necessities for the tribe, for example, work, news, stories, values, religion, and social interaction. This requires a probably more enhanced and technically complex platform than we know today, while still appearing simple from the user perspective.

No matter which of these hypotheses may come true, if any (or all), communication will become an even more important tool in the future, and the technologies to provide for communication will have to adapt to the increasing demand. We need to adapt, and adaptation means processing more data than ever before, filtering out the useless data and using the rest for control, monitoring, and decisions. It means that we need to be able to communicate on many different platforms and even be able to switch platforms during a single conversation if necessary.

In summary, Darwin would claim that to benefit from this situation, whatever that may be, we need to adapt.

Reference

1 Charles Darwin, 1871. The Descent of Man, and Selection in Relation to Sex. Available at: https://en.wikipedia.org/wiki/Origin_of_language#cite_note-35 (accessed October 13, 2016).

Index

Human Bond Communication: The Holy Grail of Holistic Communication and Immersive Experience, First Edition. Edited by Sudhir Dixit and Ramjee Prasad.
© 2017 John Wiley & Sons, Inc. Published 2017 by John Wiley & Sons, Inc.